Greening the Millennium?
The New Politics of the Environment

Edited by

Michael Jacobs

Blackwell Publishers

Copyright © The Political Quarterly Publishing Co. Ltd.

ISBN 0–631–206191–1

First published 1997

Blackwell Publishers
108 Cowley Road, Oxford, OX1 1JF, UK.

and
238 Main Street,
Malden, MA. 02148, USA.

British Library Cataloguing in Publication Data
Cataloguing in Publication data applied for

Library of Congress Cataloging in Publication Data
Cataloging in Publication data applied for.

Typeset by Joshua Associates Ltd., Oxford
Printed in Great Britain by BPC-AUP Aberdeen Ltd.

CONTENTS

Public and Parties

Lessons from Europe

Introduction: The New Politics of the Environment

MICHAEL JACOBS

What *is* 'The Environment'?

This book is about the environment as a political subject. Nowadays 'the environment' is routinely accepted as one of the principal issues of public and political concern—placed somewhere alongside (or perhaps a little behind) issues such as the economy, health, education, crime, Europe and so on. But the environment is different in striking ways from practically all of the other issues with which it is ranked. Indeed as a political subject it is in at least three ways unique.

First, it hardly counts as *one* issue at all. The simple label 'the environment' encompasses a huge range of political subjects requiring very different policy approaches and involving a wide variety of interest groups and institutions. From the impact of energy consumption on the global climate to the recycling of municipal waste; from the risks to food quality resulting from modern farming methods to the extinction of species in tropical rainforests; from the loss of countryside for housing development to the depletion of marine fish stocks; from urban traffic congestion to the preservation of wilderness; from the hazards of chemical pollution to water scarcity—'the environment' has a breadth and diversity of subject matter unmatched by any other comparably categorised political issue.

What connects these different topics is that they are all concerned in one way or another with the relationship between human society and the natural world. But this in itself does not explain why they should all be categorised under the same label. The *motivations* behind the different issues are after all very different. Anxieties about global warming, the depletion of fish stocks and water scarcity derive from questions about the long-run sustainability of economic systems—whether current trends of resource use and waste production can continue over time. Food safety, chemical pollution, urban traffic congestion and the loss of countryside, by contrast, are important because they impact directly on present health and amenity. The extinction of species and the destruction of rainforests and wilderness areas, meanwhile, raise essentially *ethical and cultural* questions about the value of the non-human world. Why are these very different kinds of issues lumped together into a single category of public concern?

The answer relates to the second source of the environment's political distinctiveness. The environment is the only 'single issue' which enters the political arena bearing its own ideology and accompanying political movement.

© The Political Quarterly Publishing Co. Ltd. 1997
Published by Blackwell Publishers, 108 Cowley Road, Oxford OX4 1JF, UK and 350 Main Street, Malden, MA 02148, USA

Education, health, taxation, crime—these may be discussed in doctrinal terms, and inspire passionate political commitment, but they don't themselves form the subject matter of an entire ideological structure (there is no 'educationalism' or 'crimism'), nor do they attract the kind of political and cultural identification which generates a social movement in its own right.[1] But this is exactly what the environment does. In its specifically 'green' forms, environmentalism is an ideology—a coherent set of beliefs about the essential nature of human society and the political principles under which it should be organised. These beliefs are held in common by significant numbers of people who identify themselves politically, and in many cases in terms of lifestyle, as 'Greens'. While the environmental movement nowadays embraces many individuals and organisations who do *not* subscribe to the green ideology in full, and would certainly not identify themselves as 'green' in an overarching sense, there can be little question that the heart of the movement—the source of its continuing momentum—remains ideologically driven. The leading organisations and activists (Friends of the Earth, Greenpeace, the anti-roads campaigners) are politically green in origin and philosophy. Even though now adopted by governments and business, the discourse of environmental policy ('sustainability', 'the precautionary principle', 'demand management', 'critical natural capital') comes directly from the conceptual frameworks and ethics of the green movement.

Indeed, it is almost certainly this which explains why the wide variety of environmental issues are treated politically as a single subject. It is the environmental movement which has achieved this. Forty years ago, before the advent of modern environmentalism, what we now perceive as 'environmental issues' (air pollution, food hygiene and public health, land development) were not categorised together at all: they sprang from different political motivations, were related to different policy fields and affected different interest groups. It is Green ideology which, seeing these issues as all symptoms of the same essential problem of human society's (over-) exploitation of the natural world, has created 'the environment' as a single political subject. The influence of environmentalism on mainstream political thought in this sense should not be underestimated.

Yet the third source of the environment's singularity as a political subject is that it is not an arena for serious disagreement between the major political parties. There are some policy differences, of course, and emphases vary, but the environment does not engender conflict between the dominant political poles.[2] Unlike issues such as the economy, education and health, where—at least in rhetoric—the parties are desperate to differentiate themselves from one another, building what are sometimes often rather minor policy differences into substantial conflicts of doctrine and worldview, the environment excites little political passion, and therefore little attempt to mark out areas of disagreement with the other side. There have been plenty of environmental news stories in recent years: the disposal of the Brent Spar oil platform, droughts and water shortages, almost-permanent protests against new roads, oil spillages off the Welsh coast, fish wars in the Irish Sea. But these have occasioned hardly any proper political argument—genuine disagreement, that is, between the major political parties and commentators over the causes of the problems or their recommended solutions. Tellingly, the most politically significant story of recent years, the BSE crisis, has barely been treated as an environmental issue at all. The 1997 general election merely confirmed this state of affairs. The environment—as in all previous general elections—simply never made it onto the agenda. The two main party manifestoes made their ritual obeisance to environmental concern, but the texts could have been swapped without anyone much noticing.

Sustainable Development

This consensus does not mean that nothing has changed. On the contrary, the mainstream view of the environment today is sharply different from what it was twenty years ago. It is simply that everyone has shifted positions together. Two decades ago environmental problems were almost universally regarded as minor, technical, soluble and politically uncontentious. They were by-products of economic growth and social progress which further applications of growth and progress would duly solve, as increasing wealth created the resources and improved technology the means to solve them. The green movement's claim that environmental problems represented a fundamental challenge to the economic system was straightforwardly rejected.

Not so now. Today few people argue in this way.[3] Throughout the industrialised world, governments and parties of both right and left now acknowledge that environmental problems are indeed very serious, requiring 'solutions' which are certainly not just technical, and may not be available at all without significant social and economic change. Few mainstream parties or governments have shown themselves willing to address these problems in any fundamental way, precisely because the solutions are so politically difficult. But in terms of basic analysis and rhetoric, the political mainstream has shifted onto the same side as the environmental movement, in stark contrast to its former oppositional stance.

The vehicle for this change has been the concept of 'sustainable development'. First introduced by the World Conservation Strategy of 1980, but popularised by the so-called Brundtland Report of 1987,[4] the concept of sustainable development has rapidly become the organising principle of modern environmentalism. With its cleverly non-specific but by no means empty definition—'development which meets the needs of the present without compromising the ability of future generations to meet their own needs'[5]—sustainable development has succeeded in overcoming the conflict between environmental protection and economic growth which characterised the environmental debate of the 1970s and early 80s. It accepts that protecting the environment requires fundamental change in the direction of economic progress and the institutions of government policy. But it argues that this is

3

compatible with continued economic growth in a (regulated) global capitalist system.

In this sense sustainable development represents a 'historic compromise' between the ideology of capitalism and its environmental critique. It has enabled a single environmental discourse to develop, used by governments (of all political complexions) and business organisations as well as by environmental groups. Allied to new scientific evidence showing the extent of global environmental damage, and institutionalised by the international agreements and policy processes arising from the 1992 'Earth Summit' in Rio, sustainable development has given a powerful impetus to environmental action in the 1990s.

But this has been achieved at a political cost. As Jonathon Porritt notes in his contribution to this volume, the environment is no longer an arena of sharp ideological battle. Of course, interpretations of sustainable development differ: business organisations and governments tend to adopt a conservative and incremental approach, seeking to balance economic and environmental goals; environmental organisations take a more radical line, arguing for environmental 'limits' and for the incorporation of social and democratic objectives.[6] But the very commonality of their discourse dampens the perception of conflict. In this sense sustainable development has been both a boon and a burden for the environmental movement. At last the environment is being taken seriously by governments and business; public commitments have been made which these bodies can be held to; new policy initiatives have been started; the pressure groups themselves have been brought into the policy process. But as a *political* subject the environment has lost its edge. Where once environmental issues dramatised a bitter conflict between opposing worldviews, now no-one fundamentally disagrees any more (or perhaps worse, they do but won't say so in public). The sympathetic hearing given even by conservative newspapers to the civil disobedience of 'Swampy' and his fellow anti-road protesters in 1997 shows just how far the mainstream political consensus has moved. Yet it is not as if this shift has been translated into fundamental political change. On the contrary, there has been little significant policy progress on any of the most important environmental issues—and no-one expects this to be any different under the new government. In some ways it seems the worst of all possible worlds: an apparently pro-environment consensus that smothers the passion of ideological conflict while in reality nothing very much changes at all.

The environment as a political subject presents, therefore, something of a paradox. Here is a set of issues which almost everyone regards as extremely serious, raising fundamental challenges for industrialised societies as they enter the new century. Opinion surveys (as Robert Worcester reveals in his chapter) show consistently high public concern, across classes and ages. The environmental movement is by far the most vibrant and visible political movement in Britain today, with close to three million members of environmental organisations, its influence now permeating cultural and social life in

4

forms such as vegetarianism, alternative medicine and green consumerism. Yet politically the environment is marginal. It fails to excite conflict between the major political parties. The Green Party is tiny and barely relevant. The environment doesn't figure in election campaigns. It doesn't even merit a Cabinet Minister in an avowedly radical new government. Why?

Relocating the Environment

The explanation offered here is that the conceptual frameworks within which environmental issues have been understood have not been adequate to the task. Though intended to illuminate the environmental problem, they have in fact merely contributed to its political marginalisation.

For a quarter of a century—from the early 1960s to the mid-1980s—the public was offered two principal ways of understanding the environment. On the one hand, as we have seen, the technocratic approach of the political and business class treated environmental issues as essentially engineering and administrative problems. Marginalisation was here part of the message. But on the other hand the green movement compounded this problem by making the environment the centrepiece of a new utopianism. Raising environmental crisis to the status of a fatal flaw in capitalist society, the green ideology of this period (and it is by no means dead) saw salvation only in the ending of economic growth and materialist consumerism. To achieve this, the entire post-Enlightenment value system of the Western world had to be overturned. It was superb critique, but hopeless politics.

The advent of 'sustainable development' has done much to bring the green movement back into the realms of feasibility; indeed on occasion the urge of environmentalists to join the political mainstream has threatened to give utopianism a good name. But for all its success in uniting former opponents on a more or less common environmental platform, sustainable development has not succeeded in making the environment into a central political issue. Indeed in some ways sustainable development seems to have combined the worst features of the old paradigms. Like the utopian tradition, the new discourse is deeply idealistic in tone, based on unashamedly value-laden statements about how the world should be organised; how government, firms, consumers and citizens should behave to achieve a more sustainable society. Its perspective sometimes seems entirely oblivious to the world as it actually *is*: to the real-life pressures and forces which dominate the modern global economy and society, and which seem in almost every case to be working *against* the ideals of sustainability and equity. Its desire to change the world seems not to be connected with the changes the world is already experiencing.

Yet at the same time the *mechanics* of sustainable development often appear to have been lifted directly from the statist managerialism of the technocrats. Sustainable development is a programme for governments to implement: devising new indicators, introducing new policy instruments, establishing

Michael Jacobs

new agencies, developing new techniques of environmental management, audit and assessment, providing information for businesses and households to encourage them to change behaviour. As Robin Grove-White argues in his chapter, though sustainable development *speaks* of empowerment and democracy, too often it just looks like the same old business of government attached to a new goal. There is little awareness here of the huge cultural changes which have been sweeping through modern societies, of the new ways in which individuals perceive their own lives and their collective identities, of the profound alienation most people feel from traditional politics and the institutions of government. On the contrary, the discourse of sustainable development seems mostly not to have noticed these changes; and in consequence to be reliant on precisely those unreformed institutions and processes from which people feel most estranged.

In these circumstances it is not really surprising that the idea of sustainable development has made so little impact on the public. Although public concern about the environment has increased markedly in the post-Brundtland period, there is little evidence that this is the product of the new conceptual framework. On the contrary, beyond the small numbers involved in so-called 'Local Agenda 21' activities at community level (about which Stephen Young writes here) the majority of people haven't even heard of it.[7] It is the peculiar achievement of sustainable development to have become the organising principle of politics for environmentalists while singularly failing to excite the interest of anyone else.

It is in this context that this book aims to offer some new conceptual frameworks for understanding the politics of the environment. An important feature of this goal is to relate environmental issues to the wider context—social, economic and cultural—in which they occur. Contrary to the impression left by the discourse of sustainable development, environmental questions are in fact centrally connected to—indeed, an important part of—the new conditions and directions of the modern world. The intention of this book is to illuminate these connections. In doing so it is hoped to relocate the environment as a subject of contemporary politics.

The book is divided into four sections, grouping together chapters concerned with different dimensions of political analysis. But the ideas of the three new conceptual frameworks run throughout. The remainder of this Introduction seeks to present these frameworks and to draw out their political implications.

The Risk Society

The first framework is that of 'risk society'. In the last ten years the theory of risk society has been widely debated amongst sociologists. In his seminal book of this name, first published in Germany in 1986, Ulrich Beck argued that Western societies were moving into a new phase of modernity or modern life.[8] The new phase is characterised by the pervasiveness of *risk*: of

6

uncertainties, insecurities and hazards. Where previously personal identity was simply a given—people were born into particular social classes, with more or less fixed value systems, family structures, communities and life-trajectories—now these dimensions of life are subject to flux and choice. Gender relations are in crisis, as both women and men are forced to relearn their social roles and remake their personal identities. Labour markets no longer offer stable, lifetime employment: not just jobs but occupations themselves may cease to exist, as new technologies are introduced and old patterns of work are changed. Life in contemporary societies has become more open-ended, less certain: there are more possibilities, but also more risks.

A central part of Beck's analysis is concerned with the environmental hazards posed by new technologies. Whereas previous forms of industrial pollution were localised and relatively easily controlled, the new technologies, such as those of nuclear power, the production of synthetic chemicals and genetic engineering, pose risks on a qualitatively different scale. These hazards are global, respecting neither national boundaries nor class divisions. They are pervasive, arising in the midst of everyday life, in foods, plastics and other materials. They are long-term and in many cases irreversible, potentially altering the life-conditions of future generations. And, though generated by scientific advance, scientific knowledge barely understands these risks at all.

The theory of risk society thus give environmental issues a central place in the analysis of the contemporary world, in two senses. At an organisational level it highlights the direction of technological advance, pointing out the central place of environmental degradation in modern economies. Beck argues that industrial societies have entered a phase of 'self-endangerment', as their environmental consequences grow increasingly large and uncontrollable. (These derive not just from new technologies but from ever-expanding use of old ones, such as fossil fuel combustion and commercial logging.) At a social-psychological level, environmental hazard is one of the dimensions of risk or uncertainty which are changing people's outlook on the world and on politics. As opinion surveys corroborate, it is a key factor in the anxiety people feel about the future, and the lack of capacity of governments to manage it.

Ulrich Beck's chapter in this book explores these themes further. He argues that science is now effectively out of control. In fields such as genetic engineering and reproductive technology, scientists are using society as a laboratory, advancing their theoretical knowledge through experimentation in the real world. The risks generated are literally incalculable: no longer simply probabilities of known events, scientists do not know what they are at all. The recent case of BSE/CJD provides a case in point. Though derided by the experts, the consumer boycott of beef was entirely sensible, since the scientists were quite unable to say how dangerous it was to eat it. But nor are these processes politically controlled. Although public authorities are held responsible for environmental disasters, the risks arise through investment decisions in research and application taken by commercial corporations. Regulatory processes lag far behind scientific advance. Moreover, lacking independent scientific resources or knowledge, politicians tend to defer to established expertise. The result, Beck argues, is a condition of 'organised irresponsibility', in which no-one is actually in control of these risks at all. So far from modern society becoming more and more bureaucratic and rationally ordered, as sociologists traditionally argued, environmental risks throw open to question and challenge established institutions, forms of knowledge and political processes.

The risk society thesis thus places the environment at the heart of contemporary politics. It makes the social control of science and technology into central political projects. It demands reform of the institutions of the state and of democracy. Beck further argues that it generates a new kind of 'sub-political' process, in which forces within civil society—interest groups, social movements and consumer activity—exert influence on and negotiate directly with corporate and administrative institutions. And he claims that it changes the nature of political conflict, as the universal impact of environmental hazards dissolves traditional class boundaries and creates shifting, issue-by-issue alliances of labour, capital and interest groups.

These arguments are taken up throughout this book, and we shall return to them below. Beck's thesis is not unchallenged, however. In his chapter Ted Benton takes issue with the idea of risk society, and the larger analysis of modernity of which it is a part. He argues that this analysis fails to understand the specifically *capitalist* logic of environmental trends. He emphasises the powerful commercial interests promoting new scientific and technological developments, and the influence of these interests over governments and regulatory institutions. He points out that many environmental risks are not actually generated by new technologies at all—BSE in fact being a prominent example, having resulted rather from the de-regulation of livestock feeds urged by agribusiness interests on a pro-free-market government. It is only by analysing the complex forces of a capitalist political economy, Benton argues, that environmental problems can be understood. In turn, it is too simplistic to argue that environmental politics transcends the traditional divisions of class or of left and right. On the contrary, he points out, environmental degradation has always hit the poorest hardest, and since the start of the industrial revolution environmental health conditions have been an important focus of labour and community protest.

Ecological Modernisation

Again, the conclusions Benton draws from this for the future of environmental politics will be taken up below. For now it may simply be noted that, though presented as competing alternatives, the underlying analyses set out by Beck and Benton need not be regarded as mutually exclusive. One can acknowledge the continuing logic of capitalism while arguing that it is now taking a particular form which is different in important ways from previous eras. So long as one's concrete analysis pays attention to the actual conditions of political economy and the character of political identity and organisation, whether one stresses the historical continuities or discontinuities is arguably of less concern.

What *is* important in this context is the recognition that environmental problems are embedded within the structures and forces of capitalist economies. This is the subject of the book's second framework.

Debate about the relationship between the environment and the economy has undergone a 180 degree turn in recent years. Twenty years ago it was almost universally believed that environmental protection and economic growth were incompatible. Industrialists and environmentalists took opposite views on which, in these circumstances, should be given priority; but they shared a common basic analysis. The dominant view today, again shared (in general) by both sides, is that it *is* after all possible to have a competitive and growing industrial economy even while resource consumption and pollution are reduced. The key to this is the efficiency with which industrial economies use the environment. The relationship between GDP growth and environmental degradation is not fixed. By using less environmentally damaging materials (such as renewable energy), improving the productivity of inputs through new technological and organisational processes (such as energy efficiency and so-called 'clean' or low-pollution technologies), and by shifting the composition of output from products with high 'material intensity' (such as manufactured goods) to those of low intensity (such as services), the environmental impact of economic activity can—at least in theory—be greatly reduced. The restructuring of industrial economies along these lines is now widely known as their *ecological modernisation.*

The concept of ecological modernisation offers a new framework for understanding the environmental-economic problem. Against the technocratic approach, it acknowledges that environmental degradation is not an incidental by-product of economic activity, which can be solved simply by add-on pollution control techniques. Environmental damage arises from many of the most fundamental features of modern industrial economies: the burning of fossil fuels, the production of food, the system of transport, the patterns of employment and service provision in cities. Changing these will mean a fundamental restructuring of economic and social organisation. But, *contra* the utopian greens, this is not incompatible with the maintenance of capitalism and continuing (if probably slower) economic growth. Capitalist economies regularly undergo structural change. Technological innovation and changing social forces stimulate the development of new production systems, land use patterns, forms of consumption and ways of life. (Witness, for example, the continuing impact of information technology and new global trading relations on the patterns of European production and consumption.) Ecological modernisation reinterprets the environmental problem as one of economic restructuring in this way: a shift onto a new path of economic development in which technological advances and social changes combine to reduce, by an order of magnitude, the environmental impacts of economic activity.

The question is whether and how this can be done within the pressures and constraints of today's global economic forces. Andrew Gouldson and Joseph Murphy examine in their chapter the potential for ecological modernisation, looking at both macroeconomic and microeconomic scales. They show that there are some connections between the ecological project and current processes of globalisation and 'post-fordist' economic restructuring. But they also reveal formidable barriers. At the level of the firm, technological innovation is an uneven and constrained process which often requires hands-on support from regulatory institutions. At the level of the economy as a whole, institutional inertia typically prevents the kind of strategic, long-term integration of policy which would promote the restructuring process.

Stephen Tindale illustrates this further in his discussion of the so-called 'environmental tax reform'. This is the proposal, widely considered central to the ecological modernisation project, that the burden of taxation should be shifted from economic 'goods' (such as labour and value added) to environmental 'bads' (such as energy, transport and waste). Tindale shows the remarkable potential of such a reform to generate both economic and environmental benefits. But he also shows the extent of opposition which the proposal has already generated. As Tindale argues, building a coalition of interests for ecological modernisation is a crucial task for the new environmental movement.

Ecological modernisation effectively re-interprets the idea of sustainable development within a broader understanding of political economy. Both concepts promote the notion that environmental and economic goals can be reconciled. But while this may be true at an economy-wide level, it does not eliminate particular conflicts. Gently puncturing any tendency to over-optimism, Susan Owens offers a subtle account of how such conflicts are played out in the real world of policy decisions. Taking the field of land use planning, where sustainable development has been enshrined in semi-statutory government guidance, she shows how different interpretations of the concept used by environmental and development interests lead inexorably to fundamental disagreements over the existence or denial of limits to growth.

Environmental Values

Owens points out that underlying these conflicts are profound questions about the nature of 'needs' and 'demands' in a rich society. Development interests argue that the forecast growth in the demand for minerals, housing, road transport and water represent the national 'need' for these resources, which the planning system must meet. Against them, environmental organisations claim that precious landscapes and habitats, good air and water quality and the control of climate change also constitute 'essential needs', and that to protect them it is necessary to 'manage' or limit economic demands. In this debate is to be found the third framework of the book.

This is that 'environmental issues' are not simply about the physical facts of damage and loss. They are vehicles which carry into the political arena the deeper anxieties and questions which many people feel about the direction and values of modern societies. This idea is made powerfully by Robin Grove-White. He argues in his chapter that the particular environmental issues which come to media and political prominence should be understood as symbols of underlying public concern. Protests over nuclear power in the 1970s reflected worries about society's increasing dependence on centralised and dangerous technologies and the capacity of the state to control them. A string of food safety scares in recent years have brought to the surface deep anxieties about the manipulation of nature in the processes of industrialised agriculture. Widespread sympathy for those protesting against the export of live veal calves and a variety of animal preservation and welfare campaigns (whales, seals, furs) express a more general concern for the ethical relationship between humankind and the natural world. These deeper issues emerge clearly in qualitative research on environmental attitudes.[9]

In this sense environmental politics cannot be understood in terms simply of the particular physical issues with which it deals. In a political climate of short-term horizons and narrowed ambitions, 'the environment' frequently acts as the bearer for the larger questions people inevitably ask about the nature of human progress. In a society as materially rich as ours, does higher output and greater income always make us better off, whatever the environmental and social costs? In an increasingly market-oriented and individualised culture, what should be the relationship between individual and society? What obligations do we have to the other species with whom we share the planet: to those we did not create, but can so easily destroy?

This theme is taken up by Michael Jacobs in his chapter on the new discourse of 'quality of life'. Jacobs argues that the concept of quality of life offers a politically feasible means of addressing the long-standing green concern about the nature of material progress and the need to reorientate Western consumption patterns. He shows how the environmental goods which contribute to quality of life (clean air, unspoilt landscapes, protected species and so on) are exemplars for a wider category of 'social goods' which are shared by the members of a society. Jacobs argues that in this way the concept of quality of life does not merely challenge the dominant view that being well off means having higher income; it also calls into question the social identity of the individual and his or her membership of the wider community.

Connecting the Environment

The argument being advanced here, then, is that the conventional frameworks in which the environment has been understood have contributed to its marginalisation in British politics. Re-interpreting the environment as a political subject in terms of the new frameworks proposed—the risk society, ecological modernisation, the environment as the bearer of wider ethical and cultural concerns—would help to bring environmental issues out of the political margins. A crucial result of this is that it would connect such issues to other important political subjects of current concern.

Thus the project of ecological modernisation links environmentalism to the wider debates now occurring over the future of industrial economies. Arguments over the processes of economic globalisation have begun to crystallise into two broad camps. There are those who see the future of European economies in terms of further deregulation, reducing social and labour market protection in order to cut costs and remain competitive in the global market place. Opposing them others argue for a new post-Keynesian settlement in which labour and social standards are maintained through the international regulation of markets and trade, and who call for greater investment in education and skills in order to maintain the advantage of European economies in high value added industries. As Gouldson and Murphy argue, advocates of ecological modernisation find themselves firmly in the latter camp. They require the protection of environmental standards from the downward pressure of trade liberalisation. And they seek active government promotion of industrial innovation and the development of new, environmentally benign industries.

In a similar way, the concept of quality of life connects environmental issues to wider arguments about the protection of shared or social goods. Over the last twenty years the dominant public philosophy of individualism and consumer choice has made it harder to defend publicly provided services, from public libraries and subsidised art to quality broadcasting. As Jacobs shows, the argument for quality of life embraces these kinds of services as well as shared environmental goods. It takes in too the crucially collective nature of personal safety and freedom from crime. Potentially, quality of life (with environmental protection at its centre) could therefore become a new organising principle for the defence of social goods and the reaffirmation of collective experience. As Jacobs argues, this could have important implications for current political debates about taxation.

The new frameworks also attach the environment firmly to concerns about poverty and social justice. As both Benton and Porritt emphasise, environmental degradation hurts the poor the most. It is the poor who suffer the worst health from sub-standard housing conditions and worsening air quality. It is the poor who rely on declining public transport services. It is the poor whose inefficient heating systems and non-existent insulation make them the likely victims of environmental taxation, unless mitigating investment measures are taken. Globally, the connection between environmental degradation and poverty is now universally accepted. As Porritt argues, the gradual (though by no means completed) acknowledgement of social justice issues has been one of the most significant developments in the environmental movement in the last decade.

Governance, Democracy and Civil Society

But perhaps the most far-reaching of the political subjects to which the new frameworks connect the environment is the question of the nature and role of the state. Much political debate over the last twenty years has centred on the decline in the power and legitimacy of government, and the need to renew the culture of democracy and the relationship between state and citizen. In various ways these questions are raised by all three of the new frameworks proposed here.

For Beck, risk society demands an opening up of the decision making process, not just of the state but of private corporations. The development of science and technology must be democratised in the widest sense. Investment strategies, scientific research programmes and the development of new technologies must all be subjected to prior public debate. Gouldson and Murphy argue that ecological modernisation will not result from the typically piecemeal and reactive character of current environmental policy making. Governments must adopt both a more integrated and a more hands-on style to encourage environmental innovation. Owens calls for institutional reform to construct a better developed 'public sphere' in which the crucial questions of value that underpin environmental conflict can be debated and judged.

This theme is taken up by Robin Grove-White. He makes more specific criticisms of the current modes of environmental governance in the UK. Grove-White argues that recent environmental controversies, such as the Brent Spar, BSE and the roadbuilding programme, have shown up the limitations of the apparently 'rational' techniques of policy making to which government is wedded. The models of 'sound science', quantified risk assessment and cost benefit analysis cannot accommodate the inescapably contested nature of environmental problems, particularly when such problems are vehicles for deeper cultural anxieties. Far from 'solving' environmental problems, reliance on such techniques is merely increasing the public mistrust of official institutions. By undermining the legitimacy of government, this will make it more difficult to promote the kind of radical regulatory programme required by the objective of ecological modernisation. Grove-White calls for a reinvigoration of citizen democracy and for more open, listening approaches to government administration.

This is, of course, easier said than done. Phyllis Starkey and Derek Osborn both respond to Grove-White's arguments—and effectively to those of Beck too. They discuss the ways in which the official institutions to which they belonged (the Biotechnology and Biological Science Research Council and the

Department of Environment respectively) have attempted to incorporate wider public debate. Starkey examines both the public's understanding of science and scientists' understanding of the public. She describes the 'consensus conference' which the BBSRC held to discuss the ethical and social issues raised by the genetic modification of plants, and suggests the possibility of a 'technology assessment' process to guide research funding. Osborn accepts that 'rational' policy making techniques are never sufficient, and defends the consultative processes which the DoE has followed in recent years in key areas of environmental policy.

Stephen Young and Bronislaw Szerszynski take the idea that environmental governance requires the rejuvenation of democracy and civil society considerably further. Young examines the range of participatory initiatives which have flourished at local level under the heading of 'Local Agenda 21'. He connects the environmental argument with wider debates about citizenship, local democracy and the development of the social economy. Szerszynski—pursuing a theme introduced by both Beck and Grove-White—argues for the crucial role played by voluntary associations in civil society. Such associations can not only implement environmental initiatives, they help to generate the public-mindedness and civic trust which is essential if environmental policy is to be successful; and they constitute a form of human flourishing—a contribution to the 'quality of life'—in their own right.

Greening the Millennium?

It was argued earlier that one of the problems facing the environment as a political subject is the lack of political—particularly party political—contestation which it inspires. One of the effects of the new frameworks offered here, it seems likely, would be to increase the intensity of political argument over environmental questions. Simply describing the issues with which the environment becomes connected suggests this strongly. In each case, the environment becomes attached to pre-existing political worldviews. Or rather, to be more precise, in each case the environment becomes attached to positions on the political centre-left.

Students of environmental politics on the European continent would not be surprised by this. There environmentalism has long been regarded as part of the constellation of left politics—by no means merely an offshoot of traditional socialist or social democratic traditions (it was in Germany, after all, that the Greens invented the slogan 'Neither left nor right but forward') but belonging in the same broad radical camp. This is one of the clearest conclusions to emerge from the surveys of environmental politics in Germany and The Netherlands offered here by Detlef Jahn and Paul Lucardie. Both stress the particularity of the national conditions which have given rise to the green parties and environmental-political circumstances in each country. But both also demonstrate the natural affinity of green politics and parties to the left, both in terms of ideology and voting support.

In the UK recognising this affinity will not be comfortable for many environmentalists. As Jonathon Porritt notes in his survey of the changing character of the environmental movement over the last two decades, one of its enduring strengths has been its opportunism and tactical flexiblity—made possible, as he notes, by its lack of ideological baggage. Porritt commends the pluralism which characterises the movement today, made up as it is now of many different interests and organisations in society, each with their own perspective on environmental change. As he observes, one of the most significant developments of the last few years has been the links made between environmental organisations seeking 'solutions' and that element of the business community which has recognised both the economic and public relations benefits of environmental commitment.

Given such a pluralism, it may seem foolish for the environmental movement to politicise itself in the ways suggested here. But this would be to ignore the new political context. Part of the reason for the environmental movement's opportunism and flexibility over the past two decades was the fact that it was faced for all this time by a basically hostile government. In these circumstances a studiously 'non-political' stance was almost certainly the only way to wield any influence at all. But the coming to power of 'New Labour' changes this. This is not because the new government is any more environmentally-minded than its predecessor (though it probably is, in a general and rather weak sense). It is because the whole ideological climate has changed. No longer do the nostrums of the free-market right constitute the political common-sense of the day. No longer do the ideas of active government, democratic reform, market regulation, public goods and social justice seem hopelessly off-beam. They can be discussed now as part of the ordinary agenda of politics. Like others with radical ideas on the centre-left, the environmental movement need no longer be on the defensive.

Indeed, the possibilities are greater than this. Labour's landslide has shifted the centre of political gravity, but the new government is hardly weighed down by ideological or even policy commitments. Beyond its first few years in government, it is by no means clear what strategic direction or ideological form New Labour will take, nor what its stance will be on some of the key issues facing the country in the slightly longer term. There is a remarkable opportunity here for the environmental movement to influence important elements of the political agenda in the new millennium. The environmental movement has things to say on a whole range of coming issues which at present few others have addressed. On employment, proposing publicly-funded jobs in energy efficiency, recycling, nature conservation and urban renewal, and questioning current practices over working time; on the quality of life in cities, linking environmental and transport policies to the reduction of inequality and crime, and an improvement in public services; on the reinvigoration of the European project, adding an 'environmental' Europe to the 'social Europe' now being developed; on food safety, the development of less chemicalised agriculture and the regulation of biotechnology; even on the millennium itself, using the century-end celebrations to focus on the theme of 'the future' and to initiate, say, a programme of global environmental protection. In all these areas, and others, the environmental movement could place itself at the heart of the new politics.

The question is, of course, how this might be done. There would appear to be two requirements. First, at least one significant environmental pressure group would have to break away from the depoliticised, diverse unity of the environmental movement as it has come to be constructed over recent years, using the new conceptual frameworks suggested here to make new political arguments and forge new alignments. The environmental movement is surely strong enough to allow this. Second, the new agenda would have to be taken up by New Labour.

Is this likely? Robert Worcester's review of opinion polls on environmental issues suggests that it would not be electorally harmful. Although environmental issues are not politically salient, this is partly because of poor party identification on the issue. He suggests that greater emphasis on environmental policy could attract considerable numbers of green-minded voters. Neil Carter's closing contribution examines the prospects. At first sight his survey of environmentalism in Britain's political parties does not look very encouraging. Though there has been steady movement in a green direction among all the major political parties in recent years—with the Liberal Democrats moving furthest—Labour's environmental commitments have remained well hidden in its policy documents, rarely venturing into its leaders' priorities.

But Carter notes that there are some signs pointing in the other direction. Labour's leadership includes some serious and long-standing environmentalists. Its ideological agenda is relatively open. And there is the Government's commitment to a referendum on proportional representation. As Jahn's and Lucardie's chapters make clear, it is a PR system which has enabled the development of green politics and parties in Germany and The Netherlands. There is no certainty that the Green Party would similarly benefit from PR in the UK, if it were to be introduced; the party is chronically weak and still largely wedded to a utopian view. But PR would surely open up the space for the wider environmental argument. Creating a more plural system, it would give strength to the growing number of environmentalists in the Labour Party and would bring the party under increased pressure from the greener Liberal Democrats. Were the referendum to lead to a change in electoral system, it *could* make the environmental difference.

Nothing in politics is certain. But the rising pressures of resource depletion and environmental damage make it extremely likely that environmental arguments will play a major role in the politics of the new century. For environmentalists—and for those who may find themselves having rapidly to become environmentalists—these promise to be interesting times.

Notes

1 It could be argued that these criteria are met by the political subjects of Europe and devolution. But one would be hard pressed to say that English nationalism (in its current anti-European form) constitutes either a fully-fledged ideology or political movement, and though Scottish and Welsh nationalism might be felt to do so, they are not in fact the driving forces behind the issue of devolution.

2 I refer here principally to the UK Labour and Conservative Parties. In the 1990s the Liberal Democrats have made considerable efforts to embrace environmentalism, and to some extent have succeeded in marking out a distinctive position. This is discussed further by Neil Carter in his chapter in this volume.

3 Those that do are sometimes known as 'contrarians', and their writings as the 'green backlash'. A fairly typical example is Wilfred Beckerman, *Small is Stupid*, London, Duckworth, 1995.

4 World Commission on Environment and Development, *Our Common Future*, Oxford, Oxford University Press, 1987. The Commission was chaired by Gro Harlem Brundtland, former Prime Minister of Norway.

5 This is the definition proposed by the Brundtland Report, and is still the most widely used. A rather more specific definition, which finds favour in many quarters, is that offered by the 1991 Report of the International Union for the Conservation of Nature, UN Environment Programme and Worldwide Fund for Nature, *Caring for the Earth* (Geneva, IUCN/UNEP/WWF, 1991): 'improving the quality of life while living within the carrying capacity of the earth's ecosystems'.

6 See Michael Jacobs, 'Sustainable Development as a Contested Concept', *Political Studies*, forthcoming. See also Susan Owens' chapter in this volume.

7 Phil Macnaghten *et al.*, *Public Perceptions and Sustainability in Lancashire*, Preston, Lancashire County Council, 1995.

8 The English translation, *Risk Society: Towards a New Modernity* (London, Sage) was not published until 1992. Beck developed the political implications of his thesis in *Ecological Politics in an Age of Risk*, Cambridge, Polity Press, 1995.

9 See for example Macnaghten *et al.*, *Public Perceptions and Sustainability in Lancashire*, *op. cit.*

Global Risk Politics

ULRICH BECK

LET us recall for a moment the intellectual situation in Europe after 1989. A world order collapsed. What an opportunity it gave to embrace a new modernity! Yet everywhere people cling to old positions, concepts, theories—and mistakes. There is now such a thing as 'left-wing protectionism'; there is even a swapping of sides in political views. Socialists have become conservative. Conservatives urge radical change; but the new world that lies behind the concepts is yet to be discovered, the script of modernity rewritten, interpreted, even invented. That is the subject of the theory of global risk society.

I would first like to sketch out the arguments of how and why the theory of a risk society can be developed as a model of a non-industrial society, and how it changes theory and politics. I would also like to take up the position of my critics by providing information about progressive arguments and conceptual ideas that help determine the development of a sociology of risk. This is useful, if only because it is better to know the critiques before one knows the theory itself. In conclusion, I would like to demonstrate some approaches—both theoretical and political—that can guide future actions, perhaps on a comparative European level.

I

Great Britain in particular, but Europe overall, is experiencing in an intense way what *The Independent* recently called 'beefgate'—the shock of living in a risk society. First of all, it makes sense to recapitulate some of the fundamental tenets of the theory of risk society. There is a big different sociologically between those who run a risk by their own decisions and those who must clean up the consequences of other people's decisions which they cannot influence. In this way, ordinary life has turned into a 'survival roulette' since the public became aware of mad cow disease. Quite banal decisions, such as whether to eat beef or not, may become life-or-death decisions. Hamlet has to be rewritten 'To beef or not to beef, that is the question'!

The arguments in the theory of risk society can be grouped into three. First, the cultural and political dynamics of a global risk society begin with the *end of nature*, i.e., the end of external risks. The threatening aspect of mad cow disease is not only that involuntary health risks are shifted onto the populace but, most of all, that these risks are not fate, but the results of decisions and options that were taken in industry, science and politics. That which seems to be an environmental problem is not a problem of the world, not an external

© The Political Quarterly Publishing Co. Ltd. 1997
Published by Blackwell Publishers, 108 Cowley Road, Oxford OX4 1JF, UK and 350 Main Street, Malden, MA 02148, USA

risk, but a number of risks that break out in the midst of everyday life and which have to be coped with by a whole series of institutions. It is part of the characteristic of such internal risks that, among other things, they are the results of efforts to *control* risk.

Secondly, the dynamics of a risk society begin with the *end of tradition*: that is, where moral milieus are replaced in the wake of advancing processes of modernisation and individualisation. On both the large and small scale, people find themselves forced to hold, or better, 'cobble', together their lives on their own initiative. The concept of risk assumes decisions and options. The more decisions, the more risks. Accordingly, the theory of risk society is closely linked to phases of individualisation in the fields of work, family, gender relationships, reflexive biography and self-identity. In that sense, the theory of risk society takes a concrete form in two special types of sociology, among others: namely the sociology of gender, the sociology of the family or pseudo-familial modes of living together; and the sociology of labour society and its erosion in the wake of long-term structural change. It sums up and comments upon the empirically well-documented symptoms of this fundamental change, which include the appearance of mass unemployment in Europe, the transformation of gender composition of the labour market, the conversion of 'normal' work relationships into 'irregular' forms of employment. The occupational system is thus being undermined by a system of multiple precarious forms of underemployment, which propagate economic insecurity right to the centre of society.

Thirdly, the theory of risk society investigates how these two connected groups have changed the epistemological and social status of science and politics. The sciences are actively tied in three contradictory ways into the advancement of risk society. Science, understood as applied technological science, causes a particular type of internal modern risk; it also specifies the language and methodological standards under which these risks are known and acknowledged. At the same time, technological science profits from these risks by building up new research fields and markets upon them. Of course, this transforms society into a laboratory in which no one organisation controls the conditions and results of any ongoing experiment.

The metamorphosis of the private sphere into risk-fraught freedom can as a consequence no longer be considered unpolitical. Accordingly, risk society turns in this sense into different arenas of subpolitics in which, for instance, the conditions for investment decisions, product development, labour arrangements and scientific undertakings and priorities are negotiated out (according to still unclear rules). Thus the image of conventional politics shifts in the political system. Society becomes 'political', though not in the sense of party politics. Rather, conflicts erupt everywhere over how we plan living and working together for the future, even where the political system with its fronts and majority relations appears ossified and obstructive.

Perhaps the enormous political and cultural power of risk perceptions can best be clarified by the currently most recent example: the globalisation risk.

We are not concerned with globalisation itself but with the rhetoric of globalisation. To use the language of insurance, the actual 'claimed loss' from globalisation is still quite limited, but the dramatisation of the possibilities is breathtaking. In 1996, for instance, the world market share of the German economy rose slightly, according to figures from the OECD. The debate in Germany on the 'business climate', however, has exposed the threat from globalisation of markets with extraordinary success. Enterprises are discovering the political potential of risk society: the sense of 'could', 'perhaps', 'if-then'.

Risk society means that the past is losing its power of determination of the present. It is being replaced by the future, that is to say, something non-existent, fictitious and constructed, as the basis for present-day action. We are talking and arguing about something that is not the case, but could happen if we do not turn the rudder immediately. Expected risks are the whip to keep the present in line. The more threatening the shadows that fall on the present because a terrible future is impending, the more believed are the headlines provoked by the dramatisation of risk today.

Globalisation has already been cited as one way to question who has power in society. Invoking the horrors of globalisation can put everything in question. This obviously applies to the power of trade unions, but equally to the sacred cows of the welfare state, the maxims of nation-state power and social insurance payments: and all of it with a gesture of regret that—unfortunately—Christian compassion must be terminated for the sake of Christian compassion. 'Globalisation' is a power drug, because believed dangers have real consequences: they stir things up.

Established risk definitions can thus wave a magic wand with which a well-fed society, nestled into the status quo, can put the fear of god into itself and thereby involuntarily and reluctantly become activated and politicised to its very core. Risk dramatisation in this sense is an antidote to the 'more of the same' blunders of the present day. A society that sees itself as a risk society is in the situation, in Catholic terms, of the sinner who confesses his sins in order to be able at least to hope for a 'better' life, more in tune with nature and the universal conscience. After all, only a few really want the rudder to be turned. Most people want to have it both ways, they want nothing to happen and they want to complain about it. Then it is possible both to enjoy the bad good life *and* any threats to it.

The theory of the global risk society also answers the question of how to understand a world which has lost the clear distinction between, for instance, nature and society. To speak of nature today is also to speak of society (culture), and if we speak of culture, then we are also speaking indirectly of nature. In this way the view of separated worlds, intimately connected to modernity's intellectual and cultural self-understanding, has long since become false, because it conceals the fact that artificial worlds are coming into existence whose peculiar feature is that they abolish or fuse together the distinctions that have hitherto been held true. This is not just the result of the industrialisation of nature. It also happens because the manufactured hazards of civilization are equally threatening to people, animals and plants. Whether we think of the ozone hole or the toxins in the air and food, whether we recall the consequences of genetic engineering or human genetic research, the picture is the same everywhere: nature has been changed, designed and endangered by human activity. One expression of this distinction is that today we no longer fear nature, but rather what we are inflicting on it. In other words, the generalisation of risk has an egalitarian effect (at least as it grows); the carefully built-up boundaries between classes, nations, people, animals and the rest of nature, between creators and creatures, are becoming porous or levelled.

Bruno Latour, in particular, has clearly demonstrated that we are living in a *hybrid* world which is abolishing the previous distinctions of modernity. The hybrid world of civilisation, which is continually being generated and changed, is the expression of equal parts of cultural perceptions, moral judgments, political decisions and technological developments. Even internal risks and hazards (co-generated by science) are man-made hybrids in this sense. That is to say, they are produced by civilisation, abolish fundamental dichotomies and offer a complex, difficult-to-decode ethical and cultural dynamic.

That last aspect can be illustrated by the fact that risks are a type of involuntary 'negative currency'. No one wants to have them or even acknowledge them; but, despite all the successful attempts to repress them, they are present and virulent everywhere. The characteristic of global risk society is a *metamorphosis of hazard*: markets collapse; there is scarcity amidst surplus; legal systems no longer fit the facts of cases; medical therapies fail; constructs of scientific rationality fall down; governments totter; swing voters walk away; everyday rules of life are stood on their heads; corporations or markets on the other side of the globe wobble or collapse. Everyone is exposed almost without protection to the threats of industrialisation. Hazards are stowaways in normal consumer life. They travel on the wind or in the water, they are concealed everywhere and in everyone, and they are passed on with the necessities of life: air, food, clothing, furnishings, through all the otherwise carefully monitored protective zones of modernity. And yet they are fundamentally dependent on our knowledge and tied into the alarms or tolerances we allow our culture. This *complex And* which resists thinking in categories of either-or, constitutes the cultural and political dynamism of risk society and makes it so difficult to comprehend. A society which perceives itself as a risk society thus becomes reflexive: that is, the foundations of its activity and its objectives themselves become the object of public controversies.

Many sociologists and social theories (including those of Foucault and the Frankfurt School of Horkheimer and Adorno) draw a picture of modern society as a technocratic prison of bureaucratic institutions and of expert knowledge. According to this image, we are all, so to speak, screws and valves in a gigantic mega-machine of technical and bureaucratic rationality. The picture of modernity that the theory of risk society draws is in stark contrast to this (and to the image that the concept of 'risk society' evidently brings with it). One of the outstanding and as yet little understood properties of the concept of risk is to *open* apparently rigid circumstances and stir them up. Different from other theories of modern societies, the theory of risk society conceives of and develops the circumstances of modernity as contingent, ambivalent and (involuntarily) open to political arrangement.

If the concept of risk becomes the leading one in culture, then there will be a break with elementary things that are usually taken for granted. Let us return to the example of mad cow disease. This risk was set off by government experts who 'cannot rule out' a connection between mad cow disease and the new syndrome fatal to humans. This inability points out that risks are often just another word for more or less egregious ignorance, for groping around in a fog of one's own making. In this way, risk society opens up a threatening space of probability. Everything falls under an imperative of avoidance. One has to bob and weave even, for instance, with the composition of menus. Everyday life turns into an involuntary lottery of misfortune. The probability of 'scoring' here may not be much higher than in the weekly football lotto, but it has become impossible not to participate in this raffle of evils. And the 'winner' gets sick or dies. Politicians such as John Major, who complained about the 'hysteria' of consumers, render outstanding service to the credibility of politics! If one takes the view that only really 'certain' knowledge demands action, one has to acknowledge that ignoring risks lets then grow immeasurably and uncontrollably. There is no better humus for risk than ignoring them.

If one chooses the opposite strategy and makes suspect knowledge (or ignorance) the foundation for action against risks, then the dams burst. Everything becomes risky. Risks only tell us what should not be done, not what should be done. To the extent risks become the only perceptual backdrop for the world, the alarm they provoke turns into paralysis: nothing works any more. Doing nothing and demanding too much can both transform the world: this could be called the *risk trap*, which is what the world could turn into with such perceptual forms of risk.

How to behave in this situation is something experts can no longer agree on. Risks which are dramatised (or hushed up) by experts simultaneously disempower them, because they force everyone to decide: what is *still* and what is *no longer* acceptable? When and where do we go to the barricades, even if only in the form of organised boycott? These questions refer to the authority of the public, the citizens, the parliaments, politics, ethics and self-organisation.

In this subversive and chaotic self-questioning ('reflexive modernisation'), something happens which sociologists faithful to Max Weber would scarcely consider possible. Max Weber's diagnosis is this: modernity turns into an iron cage in which, like the peasants of ancient Egypt, people must sacrifice, only this time to the altars of rationality. The theory of risk society elaborates the antithesis: the cage of modernity opens up. Those who are naively and realistically fascinated by a specific risk fail to recognise that it is not only

22

direct side-effects—'the toxins of the week'—which are worrying, but also the side effects of the side-effect on institutions. Not just cows go 'mad'; it also happens to governing parties, agencies, markets for meat, consumers and so on.

There is another paradox: the more we attempt to 'colonise' the future with the category of risk, the more it slips out of our control. This is an essential basis for an important distinction between two stages of the concept of risk. In a first stage (meaning in essence the beginning of the modern industrial age in the seventeenth and eighteenth centuries until the early twentieth century), risk essentially signified a form for calculating unpredictable consequences. As François Ewald argues, forms and methods for making the unpredictable predictable were developed with the calculus of risk. The respective arsenal includes statistical representation, accident probabilities, scenarios, actuarial calculations, as well as norms and agencies for preventive action. This meaning of risk refers to a world in which most things were still considered preordained, including the natural world and ways of life determined and coordinated by traditions. As nature becomes industrialised and traditions break up, new types of uncertainty arise, which Anthony Giddens and I call *manufactured uncertainties*. This type of internal risks and threats presupposes the previously mentioned triple participation of scientific experts in the roles of producer, analyst and profiteer of risk definitions. Under these conditions, many attempts to confine and control risks only expand the uncertainties and dangers.

Science

Many believe that in the age of risk there is only one authority, science. This too is based on a complete misunderstanding, not only of science, but also of the category of risk. It is the success of science, not its failures, which have called it into question and taken away its monopoly. One can even say that the more successful the sciences have been in this century, the more thoroughly they have revealed their own limitations and deficient foundations, and the more emphatically science has been transformed from a source of security into a source of reflexive insecurity. Moreover, the (technical) sciences operate with concepts of event probabilities which can never rule out the worst case.

The self-inflicted uncertainty of science applies even more to its dealings with risks, its attempt to identity these and manage the monitoring of them. In case of conflicts over risks, politicians can no longer rely on scientists. This is because they never have only a single claim or standpoint, but always several mutually contradictory ones from different agents and groups of experts, each defining risks quite differently. As has already been pointed out, this is true when the experts are good, not when they are bad. Moreover, experts can never provide anything but more or less uncertain knowledge and information on the probabilities of events; they cannot answer the question as to whether a risk is still acceptable or not. All statements on risk contain built-in standards of tolerance and acceptance relying on morality, cultural standards and perceptions, which ultimately come down to the question: how do we want to live? This is a question that can never be answered by experts alone.

If politicians then turn themselves into executive organs for scientific pronouncements and proposals, they become captives of the errors and uncertainties of scientific knowledge on how to deal with risks. This is the crucial lesson of risk society: politics and morality are (again?) gaining priority over shifting and inherently uncertain science.

Furthermore, the logic of the sciences, the very basis of their rationality, has fundamentally changed in the age of risk society. If it was formerly possible to presume a clear separation between research and theory on the one hand and technological application on the other, science has now been transformed in central fields (such as genetic engineering, human genetics, microelectronics or nuclear energy) into a technically-operating science in which the old logic of research has been turned upside-down. Previously the operative sequence was, *first*, test theories exhaustively in the laboratory and *then* evaluate and employ the results; this sequence has been reversed in the age of high-risk technological development. Nuclear reactors must first be built and used in order for their model assumptions, safety standards and so on to be tested. Test-tube babies must be 'born' in order to check experimentally the theories and assumptions on which their creation is based. Genetically altered plants and fruits must be set out and cultivated in order to check the theory behind them. The controllability of the laboratory situation has been abandoned. Society itself has become the laboratory. This raises grave consequences and issues.

Scientists have become lay people in their own fields. Before they start their research, they have no more idea than anyone else what might happen and how great the risks of their undertaking are. Yet they require the support of the public and politics in order to finance their often lengthy and expensive research projects, and this constantly forces them to assert what they cannot possibly know; namely, that they have everything 'well in hand' and that nothing really bad can happen. The philosopher of science, Karl Popper, has taught us that the foundation of scientific rationality rests above all on the ability of scientists to learn from their mistakes. In risk society, making mistakes means that nuclear reactors spring a leak or explode, test-tube babies are born deformed or people randomly fall ill or die of mad cow disease. The result is that scientists no longer dare make any errors. They *do* make errors, however, and these have more and more far-reaching consequences. Ultimately, scientists inquire into and study one another's mistakes, results and the dangers they unleash more and more keenly.

Not only has society become a laboratory, there is also no longer anyone who could be made responsible for the results. Experiments such as nuclear power generation or biotechnologies cannot be confined in terms of time, space or number of affected persons. At the same time, there is no longer anyone who monitors the experiment and no one who ultimately makes scientifically-founded decisions on the validity of the initial hypothesis.

What role does politics play in this context? Actually no direct decisions on the development of technologies (apart from the peaceful use of nuclear energy) are made in the political system. On the other hand, if anything goes astray, the political agents and institutions are held responsible for decisions they did not really make and for consequences and hazards of which they have no justified notion.

Industry is in a doubly favoured position in its relationship to governmental politics and Parliament; first, because of the autonomy of private investment decisions and, secondly, because of its monopoly on employing and applying technology. Politicians are in a correspondingly disadvantaged position. First of all, they struggle to keep up with what is now being hatched out in engineering laboratories. Studies show that most parliamentarians obtain their information on technological developments almost exclusively from the mass media. Despite all the funding to support scientific and technological research, the influence of politics on the goals of technological development remains secondary. Ultimately, no decisions are made in Parliament on the application or non-application of microelectronics, genetic technology and the like, or on the goals of their development. In the great majority of cases, the parliamentarians decide across party lines exclusively to give support and to accelerate technological developments in order to protect the future of the economy, especially jobs.

That is to say, the separation of powers cedes to industry the right to make decisions without a corresponding responsibility for the risks those decisions unleash for the public, while politics is assigned the task of democratically legitimating decisions it never really made and knows little about. In the worst case, a threatening or real catastrophe, politicians must justify and often vouch with their credibility for decisions which were made elsewhere by others.

The Principles of Calculability

The consequence is that no one is responsible for risks. Neurotechnologies and genetic engineering rewrite laws that have thus far determined human thought and life. Who exactly does this? Scientists? Politicians? Industrialists? The public? No matter who one asks, the answer is: no-one. Risk politics relies on the 'faceless power' that Hannah Arendt considers the most tyrannical form of the exercise of power, because under these conditions, no one can be held responsible for anything. Sometimes, in the spotlight of potential or actual catastrophes revealed by the mass media, bureaucracies suddenly stand there naked, and the alarmed public becomes aware of what those agencies really are: forms of organised irresponsibility.

This is by no means everywhere and always the case. If one goes back to the distinction between external and internal risks, then early industrial societies developed rules and institutions to make the unforeseen and unintended consequences of risks 'controllable'. The welfare state stands for a model which answers the question of how spatially, temporally and socially *limited* risks can be answered collectively and in an institutionalised manner: namely, by rule-directed attribution of guilt and liability, legal norms for compensation, balanced actuarial principles and collectively divided responsibility. The classical example for this is the development of insurance contracts in cases of accident, injury, destruction, unemployment, etc.

Yet the outstanding characteristic of risk society is that these precautionary norms and insurance regulations, according to which causalities and costs are distributed, have been abolished or undermined by central industrial and technological developments such as nuclear energy, biotechnologies, human genetics and the like. Risk society means a balancing act beyond the limit of insurance, indeed, insurability. It can even be said that the amount of insurance protection diminishes with the size of the danger, and this is only one symbolic indicator that global risk society operates beyond the principles of calculability which apply in institutions. Compared to the possibilities of assigning guilt, liability and costs which classical (primary) modernity had available, global risk society—the second modernity—has no securities and guarantees of that type.

Anyone who inquires into the actual basis of the *political* alarms over the 'ecological crisis', can derive the following answer from the theory of risk society: turned political, the ecological crisis amounts to a *systematic infringement of fundamental rights*. Ultimately, this is a crisis of fundamental rights whose long-term effect, the vitiation of social legitimacy and political power, can hardly be overestimated. Hazards are created by industries, externalised by business, individualised by the legal system and trivialised by politics. The vacuum of power and legitimation created in this way manifests itself only if the system and its agents are tested concretely, as in the actions of Greenpeace.

II

Having thus traced several lines of argument that I worked out in the theory of the global risk society, I would now like to change sides and take up some of the criticisms and objections that this theory has experienced. In a scholarly conference in Great Britain a few months ago, Professor Hilary Rose said, 'I have the impression that there are specifically German background assumptions and levels of meaning in the theory of risk society', and she added, 'Perhaps Great Britain cannot afford to be a risk society'. In her understanding, the theory of risk society has a note of wealth and security typical of post-war Germany. Or could one say, taking up an objection from France, that risk society means a society obsessed by 'le Waldsterben' [the destruction of forests by pollution]?

Perhaps that is the case. Certainly this theory is one of the few attempts to open the social scientists to ecological issues. In fact, being Green has virtually

26

become a part of German post-war identity. Many Germans would like to have a Germany that was like a larger, Green version of Switzerland. They claim to be or hope to become the ecological conscience for the world. On the other hand, it may be that the testing of atomic bombs is part of the national identity of France. This makes it clear that risk conflicts are not just intercultural conflicts, but conflicts in which 'contradictory certainties' clash. People, experts, cultures, industries, classes, regions and nations, none of whom want such conflicts or this trans-border risk situation, are nonetheless drawn into them. This is not happening only in the abstract (legally, economically and politically), but also in the conduct of everyday private activities. Perhaps it is not so wrong to say that something like a European public has come into existence unintentionally and involuntarily in the wake of the conflict over 'British beef'. Perhaps mad cow disease in Europe, putting everyone into conflict with everyone else, brought about a kind of negative European identity. Even in small village butcher shops and inns in Bavaria, one suddenly finds confidence-building measures intended to put people on personal terms with the animal whose flesh is about to be consumed. The farmer and his family, who fed and cared for the animal in their stalls and kept it healthy until it was ready to eat, offer their regards, complete with photo. The Europeans, about to burn a hundred thousand head of cattle in order to stabilise the market for meat, appear completely mad to people in India. Devout Hindus offer to care for the doomed sacred cows in Britain until they meet their end. By the way, mad cow disease broke out as an issue in Britain only a few weeks after the conference at which Hilary Rose said Great Britain could not afford to be a risk society.

Reflexive Modernisation

Here it becomes clear that it is important in the case of risks to distinguish between *knowledge* and *results*. Not wanting to know is an essential precondition for the growth of risks (see above). In that sense, one can distinguish two phases in risk society. First there is the stage in which the consequences of self-endangerment are systematically generated but do not become public issues or the centre of political conflicts. What predominates here is the self-concept of industrial society which, under the rubric 'residential risks', simultaneously multiplies and legitimates the hazards produced as a result of decisions (the 'residual risk society').

A completely different situation arises when the hazards of industrial society begin to dominate public, political and private debates and conflicts. Here the institutions of industrial society turn into producers and legitimators of hazards they cannot control. This transition occurs while property and power circumstances remain unchanged. Industrial society views and criticises itself as a risk society. Society still makes decisions and operates according to the pattern of the old industrial society, while the interest organisations, the judicial system and politics are already overgrown with debates and conflicts arising from the dynamics of risk society.

In light of these two stages and their sequence, it is possible to introduce the concept of 'reflexive modernisation'. This does not mean (as the adjective 'reflexive' might suggest) reflection, but rather self-confrontation and self-transformation. The transition from the industrial to the risk epoch of modernity occurs *un*intended and *un*seen, inexorably in the wake of the autonomous dynamics of modernisation, following the pattern of unintended consequences. One can say that the constellations of risk society are created because the certainties of industrial society (the consensus on progress and the abstraction of ecological effects and hazards) dominate the thought and action of people and institutions. Risk society is not an option that is selected or rejected as the result of a political conflict. It originates in the inexorable motion of autonomous modernisation processes that are blind to their consequences and deaf to their hazards. Taken altogether, the latter produce latent self-endangerment that calls the foundations of industrial society itself into question.

In all of my books, I attempt to show that the return to the theoretical and political philosophy of industrial modernity in the age of global risks is doomed to fail. Such orthodox theories and policies remain tied into false hopes and assumptions on the automatism of 'progress' and the inherent necessity of technological developments. The prevailing misconception is that we can get a grip on the risks confronting us with the methods and models of the nineteenth century, which were built on temporally, spatially and socially *limited* accidents. Quite similarly, many believe that the disintegrated institutions of industrial modernity—the nuclear family, stable labour markets with full employment, the separation of genders and social classes—can be reinvigorated in order to build up dams against the storm floods of reflexive modernisation currently assailing the West. Attempting to apply the ideas and categories of the nineteenth century to the twenty-first century is the prime error of sociological theory, the social sciences and politics. It may therefore make sense to state this more pointedly and thus both sharpen and focus upon central theoretical concepts which have not really been picked up, discussed, and advanced or rejected in previous debates. In all brevity I would like to go into three crucial concepts: 'organised irresponsibility', 'relations of definition' and the 'social explosiveness of hazards'.

The concept of 'organised irresponsibility' is intended to explain how and why the institutions of modern society must acknowledge potential and actual disasters, while simultaneously denying their existence, concealing their causes and ruling out compensation or control. This can be demonstrated with a paradox which should be publicly discussed. We are faced with an advancement of environmental destruction (threatening or actual) *and* an enormous growth of environmental laws and regulating institutions on both the national and the international scale. At the same time, the laws of *faceless power* apply everywhere (see above). How is this possible? The explanation lies, I think, in the mismatch which exists in the epoch of global risk society between the character of the internal hazards generated by late industrialism and the prevailing 'relations of definition', which relate by origin and content to an earlier qualitatively different period.

This concept of 'relations of definition' is conceived as a parallel to Karl Marx's concept of the relations of production, specifically in the global risk society. It refers to rules, institutions and resources which determine the identification and definition of risks. This is the legal, epistemological and cultural matrix in which risk policy is organised and practised.

What it means in detail can be outlined by four questions. (1) Who—that is, what social agency and authority—establishes in what way how harmless or dangerous products and their side-effects are? Does the responsibility lie with those who create and profit from the risks, or with those who are currently or potentially affected, or with public agencies? (2) What type of knowledge or unawareness of causes, dimensions, agents and so on is consulted or acknowledged here? Who bears the burden of proof? (3) What is considered 'sufficient proof' And this, of course, must be answered in a world where all knowledge of hazards and risk moves in the presuppositions of probability theory. (4) Where hazards and destruction are recognised and acknowledged, who decides issues of liability, compensation and costs for the affected parties, and who rules on appropriate forms of future monitoring and regulation?

If one thinks these questions through, it becomes clear that in all these respects, risk societies are caught and biased by semantics and models built into *the relations of* definitions of industrial modernity which minimise or conceal the latter's continual production of hazards. The relations of definition prevailing in the law, in science and in industry are not only completely unsuited to the (creeping) disasters set into motion and maintained by industrialisation, they are also inappropriate for the general insecurities that those disasters provoke. Accordingly, we confront the paradox that the hazards grow and become more obvious, and yet they simultaneously slip through all the nets of evidence provision, accountability and liability with which the legal and political institutions attempt to confine them.

There is the same question everywhere: who or where is the political subject of global risk society? I have gone to great trouble to find an answer to this question. But the one I have proposed has not yet been perceived, theoretically or politically. My argument, crudely oversimplified, is that at the same time everyone and no one is the potential political subject of risk society. It is perhaps not surprising that this answer was ignored. My theory is not exhausted by pointing that out, however. Instead, I have in mind—similar to the theory of quasi-objects propounded by Bruno Latour—that, to the extent they are perceived publicly, hazards become quasi-agents. That is the meaning of the metaphor of the *social and political explosiveness of hazards*.

This explosiveness is created by the contradictions in which hazards entangle monitoring (administering) institutions in risk society. I attempt to show how the public perception of large-scale hazards and risks ignites it own dynamic of cultural and political change which undermines the governmental bureaucracies, calls into question the claims of scientists and experts to be authorities and reshuffles the established antagonisms and political parties in the political sphere. The quasi-activity of hazard, which breaks out of its social origin against the background of organised irresponsibility and antiquated relations of definition only when those hazards are illuminated by the mass media, may be activated or used by social movements. Certainly, Greenpeace, among others, has been successful in harnessing and bringing into focus their latent dangers.

On the other hand, governments and bureaucracies, as well as experts and industries, have polished routines for concealing and denying hazards. Data may be withheld, not gathered in the first place, or called into doubt for methodological reasons. Opposing experts and arguments are mobilised. The expert system in general can be fortified from the inside against any self-criticism. Acceptable levels can be corrected upwards or never set in the first place. In the worst case, it turns out that 'human error' and not systemic risk was to blame, which rules out any pressure for change in advance. Nonetheless, at least from the viewpoint of the global risk society, battles are being fought here in which victories are always only tentative and defeats almost probable, precisely because answers from the nineteenth century are being projected onto the challenge of global risk society. Moreover, this can be demonstrated by anyone at any time.

Differential Politics

There have been two major objections to these concepts. First, it is said that they underestimate or ignore the significance of social movements; secondly, they are said to be based on conceptions of a welfare or insurance state that universalises the German case and does not apply, for instance, to the USA, Great Britain or the Scandinavian countries.

The theory of the explosiveness of hazard is not intended by any means to be a new type of automatism. On the contrary, it provides political movements with the external conditions, strategies and resources for a type of 'judo politics'. That is to say, it permits them to unleash powerful impulses for change from positions of relative powerlessness, to 'shape' them and make them permanent. Special preconditions must be present for this politics of oppositional power and monitoring to be activated. The first of these is a strong and independent judicial system, the second is a relatively free and critical mass-media, and the third is a widespread willingness to engage in social self-criticism, as is possible through the connection of consumer society and direct democracy in the form of a mass boycott. To cite one example, this was put into practice across cultural and national boundaries in the conflict over the disposal of the Brent Spar oil-drilling platform. This could be a model of a *different politics*, in which the subpolitical agents can obtain the opportunity to exercise power.

It is certainly true that the political explosiveness of hazard presumes the explicit responsibility of the state for matters of safety and security. It is equally true that decentralising trends are now aiming to relieve the over-burdened state from more and more of its tasks, including precautions against hazards and risks. Yet, even in deregulated countries like Great Britain, ministers or their representatives must still be accountable to the public when disaster strikes. If an oil tanker runs aground off the Welsh coast, for instance, and contaminates for years to come miles of coastal landscape living off tourism, people do not initially question the owner of this tanker (which is generally hard to clarify, since it sails under different national flags). Instead, some representative or agent of the government must answer the questions of an anxious public regarding the extent of damage and the amount of loss or compensation they face, even though he does not have jurisdiction over this situation according to the rules of the institutional game.

This governmental obligation to provide safety, which is certainly particu-larly highly developed in continental European welfare states, forms the background for the social explosiveness of hazard even in the deregulated capitalist states of the Anglo-American type. One need only contemplate the issues and follow-on problems of the privatisation of nuclear power plants and their security provisions to realise that deregulation may even widen the gap between the government's safety claims and its actual lack of responsi-bility, and thus deepen the state of organised irresponsibility.

III

I would like to point out two conclusions in closing.

First, the theory of global risk society—as it is hoped will have become clear—is not concerned with exploding nuclear submarines that turn up somewhere. Neither is it, as some would like to assume, just one more expression of 'German angst' as we approach the millennium. On the contrary, it is an attempt to develop a model for the present which will help towards a fresh understanding of the future. My argument interprets a theory of postmodernism as the stage of a radicalised modernity, a stage in which the dynamics of individualisation, globalisation and risk destroy the foundations of primary modernity and brings into existence a second, unknown modernity, which is yet to be discovered. 'Inside the West', writes Gottfried Benn in his famous Berlin letter of 1948, 'the same group of intellectuals has been discussing the same group of problems with the same group of arguments relying on the same group of causal and conditional clauses and has been reaching the same group of results, which they call syntheses, or of non-results, which they call crises—the whole things seems a bit tired, like a popular libretto. It seems rigid and scholastic, it seems like a genre play relying on backdrop and straw'.

ULRICH BECK

A New Democracy

The theory of global risk society attempts to break out of this intellectual matrix and bring things to consciousness that were conventionally considered 'excluded middles'. This comes down to a peculiar reversal of the Feuerbach-Marx controversy. *Thinking* must be changed so that the world of modernity can be renewed with its own origins and demands. Only the power of concepts can open the political space to reforms, not to mention an (ecological) reformation of the Western symbiosis of capitalism and democracy, which only *appears* to be eternal. We must learn to *see* that the fatalisms which dominate our thinking are antiquated and can no longer withstand the test of a resolute self-confrontation of modernity. In the mouths of philosophers and sociologists, the term 'rationality' has generally come to mean 'discourse' and 'cultural relativism'. My theory of 'reflexive modernisation' implies, by contrast, that we do *not have enough* rationality and reason ('*Vernunft*').

Secondly, the outstanding characteristic of the concept of risk, as conceived of and elaborated here, has not yet been sufficiently appreciated, theoretically or politically. It lies in the fact that previously depoliticised fields of decision-making are becoming politicised by the public perception of risk. Generally involuntarily and against the resistance of the powerful institutions monopolising these decisions, they are opening up these problems to public doubt and debates. In global risk society, therefore, subjects and themes once treated behind closed doors, such as public investment decisions, the chemical composition of products and medicines, scientific research programmes and the development of new technologies, are articulated and debated in public. All these must justify themselves until it becomes possible to work out and implement a legal and institutional framework to legitimate and preserve this important extra bit of democracy. Unintended and unseen 'secondary problems' are being debated everywhere as if by pre-arrangement, even before the relevant products and technologies have actually been invented. The crucial point, of course, is that this expansion and deepening of democracy into the apolitical regions of business, economics and science (and to some extent into the private sphere as well) was blocked until now by antiquated relations of definition which placed the burden of proof on those who were harmed by risks rather than on those who profit from them.

In global risk society, it is not merely true that a self-reflexive or self-critical society is coming into being as a result of public self-perception within the framework of reference to risk. The contours of a *utopia of ecological democracy*, which means for me in essence a *responsible modernity*, are also coming into view. This creates an important addition to the debate in Anglo-American philosophy and theory of science which is being conducted under the key words 'technological citizenship'. These discourses lay out the image of a society that debates the consequences of technological and economic development *before* the crucial decisions are made. The burden of proof for future

32

risks and hazards would lie on those creating them, and no longer on those potentially or really threatened or injured. This is a change from the *polluter-pays* to the *polluter-proves* principle regarding the possible harm plans may cause. In particular, it would be necessary in science and the law to find or invent a new system of rules that redefines, negotiates and establishes what constitutes 'proof', 'adequacy', 'truth' or 'justice' in view of probable global risks and hazards. What is needed, therefore, is nothing less than a *'second modernity'* with which our eyes, our understanding and our institutions could be opened to the self-imposed immaturity of primary industrial civilisation and its self-endangerment.

Many theorists and politicians overlook the *opportunity* of global risk society to which I would like to draw attention with this theory. Thus the European debate on mad cow disease stirred up important issues on, for instance, the role of self-imposed unawareness in risk conflicts. The fact that the genetic engineering industries, following the political and economic disaster of nuclear energy, are now being forced to justify themselves publicly and defend their plans to the searching eyes of a worried public, is also one of the achievements that place global risk society, despite its name, in a by no means hopeless light. This example, of course, shows once again how the established relations of definition codify the prevailing organised irresponsibility and cut off any incipient public debate over the future and further development of Western democracy.

This might be precisely one of the topics that should be investigated in the future by social scientists, if at all possible, on a comparative European level. The object of such research would be to explore how the social definition of risks and risk management fails under varying cultural conditions, and how it can be reconstructed in order to uncover the peculiar logic and (negative) power of conflicts over and definitions of risk. In the latter, after all, people and cultures which do not intend to have anything to do with one another (at least in matters of risk) are forced together. This is known to some extent and is already happening, but working out and focusing research projects of this type, in which questions of organised irresponsibility and antiquated relations of definition in different European countries occur, could be the beginning of a genuinely new and cooperative adventure in research.

Biographical Note

Ulrich Beck is Professor of Sociology, University of Munich. His books include *Risk Society* (1992), *Ecological Politics in an Age of Risk* (1995), *The Renaissance of Politics* (1996), *The Normal Chaos of Love* (1997) and, with Anthony Giddens and Scot Lash, *Reflexive Modernisation* (1994).

Beyond Left and Right? Ecological Politics, Capitalism and Modernity

TED BENTON

Introduction

Fears of impending global ecological catastrophe pervaded public debate in the early 1970s. The very rapid growth of environmental social movements since that time has ensured that the question of human impacts on our natural environment has remained on the public mind. However, the bundle of issues which have come together to make up the 'environmental agenda' are really quite diverse and heterogeneous, as are the perceptions and strategies of the groupings which make up the 'environmental movement'. The catch-all discourse of 'sustainable development' has provided the openings through which formerly quite radical environmental organisations have acquired respectability and 'insider' status in some spheres of governmental and inter-governmental policy-formation. At the same time, however, radical ecology movements and groupings continue with their 'outsider' role as an oppositional force in society. Sometimes, as in recent road protests in the UK, they are able to capture significant and sympathetic media coverage on the basis of imaginative and courageous forms of direct action.

There is little doubt that the emergence of radical social movements concerned with ecological issues is a new and significant phenomenon. However, social scientists are divided over just what this new phenomenon means for our understanding of contemporary society, and for the future of radical politics itself. Perhaps currently the most influential view is that offered by a group of sociologists, of whom Ulrich Beck and Anthony Giddens are the best known, concerned with the analysis of 'modernity', or the modern age. For them, the new environmental politics constitutes an emergent kind of radicalism which displaces and goes beyond the old political divisions of left and right. It is the form of politics associated with a new phase of modernity, a phase they call 'reflexive modernisation'.

According to the theorists of 'reflexive modernisation', the institutional and cultural forms and the major sources of social and political identity and conflict which characterised earlier phases of modernity, are in the process of being displaced. One of the key features of the new phase is the significance of ecological destruction and large-scale hazards in the trans-formation not only of the physical, but of the moral and political landscape. The changes in modernity and the associated emergence of new social movements, they argue, holds out the prospect of a 'new' form of modernity, both democratised and sustainable. Implicitly or explicitly, the

Published by Blackwell Publishers, 108 Cowley Road, Oxford OX4 1JF, UK and 350 Main Street, Malden, MA 02148, USA

writers foresee in this process the demise of recognisably leftist and rightist politics.

Whilst I have considerable sympathy with the normative stance of these writers, and indeed for their project of fully incorporating the socio-ecological dimension into social theory, I shall be arguing that the analytical concepts they use to explain the rise of ecological politics are deeply flawed. This has important implications for their views on the future of radical politics. The key difference between the position I shall be advocating and the 'reflexive modernisation' school is in the rival frameworks of ideas through which the present historical period is understood.

Reflexive Modernisation and the Politics of Risk

For the advocates of reflexive modernisation, history is understood as a linear sequence of stages, from traditional (or pre-modern) society through a 'simple' form of modernity to 'reflexive' modernity. Modernity is char-acterised in terms of a list of institutional forms or 'dimensions', none of which is given overall causal priority.

For Giddens, simple modernity (essentially, western society since the Enlightenment) is characterised by four 'institutional' dimensions: political/administrative power, typically liberal, representative democracy, an eco-nomic order overwhelmingly capitalist in form, with the now defunct communist regimes as a temporary variant; a relation to nature defined by modern science and industrial technology; and a state monopoly on the legitimate use of violence. Giddens argues that conservative, liberal and socialist traditions in politics are connected to this phase of modernisation, but are now exhausted as a consequence of changes to modernity occurring over the last 40–50 years.[2] These processes are summarised by Giddens as 'globalisation', 'detraditionalisation', and 'social reflexivity'. Giddens resists an economic account of globalisation, focusing instead on the implications of new communications technologies and mass transportation. Partly because of the cultural cosmopolitanism which flows from globalisation, traditions which have persisted into, or become established during simple modernisa-tion, can no longer be justified or legitimated 'in the traditional way': they have to justify themselves in the face of alternatives. Individuals no longer have their lives set out for them by the contingencies of their birth, but are constantly faced with choices about how to live: whether to have children, how to dress, what to believe in and so on. The establishment of personal identity, in other words, increasingly becomes a life-project of 'reflexive' individuals.

The newly emergent conditions of reflexive modernisation, according to Giddens, render the inherited political traditions obsolete. Traditional forms of class-identity are dissolved; changes in the labour market, in gender relations and family structure render the institutions of the welfare state unsustainable and inappropriate; globalisation and reflexivity in consump-

tion and life style choice make centralised forms of economic control unworkable. The established parties and political institutions consequently lose their legitimacy. However, this is not the end of politics—nor even of radical politics. Drawing on a schematic account of the new social movements as forms of resistance to each of the institutional dimensions of modernity, Giddens postulates the emergence of a radical 'generative' or 'life' politics beyond the old polarities of left and right.[3]

In response to the political/administrative system, there are social movements aiming at the radicalisation of democracy, and against surveillance and authoritarianism. Reflexive modernisation also involves democratisation of personal life, in which relationships between lovers, friends, parents and children and so on are no longer governed by traditional assumptions and expectations. In the sphere of capitalist economic relations, social polarisation and fragmentation continue to characterise reflexively modernising societies, but the supposed demise of class politics leads Giddens to suggest (rather vaguely) that these problems will be corrected by a post-scarcity order which owes as much to ecology and conservatism as it does to socialism. In the dimension of science and industrial technology, the power of simple modernity to control the forces of nature has generated a new order of risk—'manufactured' risk—to which the green movement has responded with a utopian desire for a return to authentic nature. In the dimension of institutional violence, the peace movement points to a growing role for dialogic forms of conflict-resolution in a post-traditional, reflexive world.

The German sociologist Ulrich Beck has a great deal in common with Giddens's way of thinking, but has a much more highly developed and original approach to the ecological dimension of reflexive modernisation. In his view, the processes of detraditionalisation, globalisation and reflexivity (which he defines rather differently from Giddens) are leading to the emergence of a new stage of modernity which deserves the title 'Risk Society'.[4] Risk and uncertainty increasingly pervade all dimensions of personal and social life: increased rates of divorce and family breakdown, uncertainty and vulnerability in the labour market, and most centrally, for Beck, uncertainty in the face of the hazards generated by new, large-scale industrial technologies and by advances in scientific knowledge.

Beck shares with Giddens a historical periodisation of risks and hazards. In pre-modern times, risks (in the shape of disease epidemics, floods, famine and so on) were experienced as having an external source, in nature. Simple modernisation, with the development of industrial technologies, displaced 'external' risks in favour of self-created, or 'manufactured' risks, by-products of industrialisation itself. In Beck's view, reflexive modernisation ushers in a new order of manufactured risk with profound cultural and political implications. The 'semi-autonomous' development of science and technology unleashed under simple modernisation has through its own dynamic yielded new large scale technologies in the nuclear, chemical and genetics industries which pose qualitatively new hazards. They put modernity itself at risk.

36

The hazards of risk society are qualitatively different from those of simple modernity in at least seven ways: (1) They are unlimited in time and space, with global self-annihilation now a possibility. (2) They are socially unlimited in scope—potentially everyone is at risk. (3) They may be minimised, but not eliminated, so that risk has to be measured in terms of probability. An improbable event can still happen. (4) They are irreversible. (5) They have diverse sources, so that traditional methods of assigning responsibility do not work. Beck calls this 'organised irresponsibility'. (6) They are on such a scale that they exceed the capacities of state or private organisations to provide insurance against them or compensation. They may be literally incalculable. (7) They may only be identified and measured by scientific means. Consequently the contesting of scientific knowledge and growing public scepticism about science are important aspects of the 'reflexivity' of the risk society.

The pervasiveness of risk, and especially of the new order of hazards generated by large scale technologies, leads Beck, like Giddens, to see reflexive modernisation as a political watershed. In Beck's work, two clusters of themes are prominent. The first is the supposed demise of class conflict over the distribution of goods. Beck takes class conflict between capital and labour to be characteristic of the period of simple modernisation. But this is in the process of being displaced both by a new agenda of political issues and by new patterns of coalition and conflict. Giddens and Beck agree that severe material inequalities continue to exist through reflexive modernisation. But for them, globalisation, detraditionalisation and reflexivity erode traditional forms of class consciousness and identity, so that class relations are increasingly individualised and conditions for collective class action disappear. In this respect Giddens and Beck are in line with a welter of recent announcements of the 'death of class'.[5] Beck's particular gloss on this thesis includes the claim that the political agenda is undergoing a shift from conflict over the (class) distribution of goods to conflict over the (non-class) distribution of 'bads', that is, the environmental costs of continuing industrial and technical development. The new patterns of conflict characteristic of the risk society will involve shifting patterns of coalition and division defined by the incidence of these costs. So we can expect workers and managements in environmentally polluting industries, for example, to be in alliance with one another against those in industries such as, say, fisheries or tourism, which suffer from pollution. Finally, there is the implication that the new order of environmental hazards constitutes the basis for a potentially universal interest in environmental regulation, since the relatively wealthy and powerful can no longer avoid these hazards, in the way they could escape the risks associated with earlier industrial technologies.

The second cluster of themes marking a qualitative break with the politics of the past is also centrally connected with environmental hazards. Here it is the challenge these hazards pose to the steering capacity of modern states, and so to political legitimacy. In essence, Beck's argument is that under simple modernisation, legitimacy was achieved through the progressive development of a welfare/security state, in which either public or private institutions provided guarantees against risk in the various dimensions of life—public health care provision, pensions, unemployment and sick pay, welfare benefits and so on. Reflexive modernisation, characterised by changed gender relations, family breakdown, flexible labour markets, and, above all, hazards of unprecedented scale and incalculability, exposes the growing inadequacy of the welfare/security system to deliver what it has promised. Beck proposes the notion of a 'sub-politics' which might emerge in response to this situation, as social forces in civil society press for more democratic participation in decisions currently taken by hierarchies of technocrats and top business executives.[6]

It seems to me that the theorists of environmental politics in the light of reflexive modernisation successfully allude to significant changes in contemporary societies. Their characterisations of these changes are often imaginative and persuasive. However, I want to show that in several key respects their claims are either empirically mistaken or theoretically defective, or both. An alternative, ecologically informed analysis of capitalist economies and society is more adequate, I shall argue, to the explanatory task at hand, and points to quite different political possibilities.

Capitalism or Modernity?

In their understandable anxiety to avoid economically reductionist accounts of the relationship between capitalism and other fields of social life, the reflexive modernists are reluctant to assign any causal significance to the economy beyond its own boundaries. We are left with a typology of institutional 'dimensions' of 'modernity', but with no attempt to characterise the relationships between them, nor the processes through which they constitute linked elements of a whole society. The reader is given the impression that each dimension is to be understood as an autonomous causal order in its own right. The same applies to the social movements which, in Giddens's account, arise as forms of resistance to each institutional dimension. Though he recognises the anomalous character of the women's movement as transcending his institutional divisions, he continues to see, for example, the environmental and labour movements as separated from each other in their resistance to distinct institutional dimensions.

But one doesn't have to be an economic reductionist to acknowledge the causal links between capitalism and the institutions of society and the state. It seems clear, for example, that European governments will be limited in their capacity to alter unsustainable agricultural regimes, or to reduce carbon dioxide emissions, or to constrain motor traffic growth (and so on) for basic reasons which have to do with the economic power of capital. In part, this power takes the form of extensive 'colonisation' of government ministries and EU institutions by organised agribusiness, energy producers, road transport interests and other industrial groups. In part it results from the dependence of state economic policy itself on the profitability of the relevant industrial sectors. To point to such connections is not to be committed to economic determinism, but, rather, to suggest that it is implausible to assume *a priori* that each institutional dimension can be understood independently of its structural ties with the others. Similarly, while Giddens and Beck rightly acknowledge that capitalist development continues to generate material inequalities (in fact, the evidence is of *increasing* polarisation of wealth and power as a result of economic globalisation and deregulation) they give accounts of their favoured 'life-' and 'sub-' politics in ways which seem largely innocent of the implications of these economic inequalities for the other 'dimensions' of modern society—for the conduct of private lives, for the prospects of democratisation in the political system, or for the political potential of social movements.

This unwillingness to acknowledge the causal influence of capitalist relations and forces undermines the reflexive modernisers' account of the development of science and technology. These are seen as autonomous processes. In the case of Giddens they constitute a distinct institutional 'dimension' separate from capitalist economic relations. In Beck's work a parallel separation is achieved by his identification of science and technology as expressions of an abstractly defined 'instrumental reason', the legacy of the European Enlightenment, and characterising the distinctively modern relationship to nature. The upshot for both writers is a portrayal of ecological crisis and large-scale hazards as consequences of a secular process of scientific and technical development, endemic to a definite stage of 'modernisation', and subject to resistance on the part of single-issue environmental movements. These latter are then somewhat condescendingly criticised for their utopian and retrogressive desire to return to an 'authentic' nature which, apparently, no longer exists. ('Today, now that it no longer exists, nature is being rediscovered, pampered. The ecology movement has fallen prey to a naturalistic misapprehension of itself. . . .').[7]

What is wrong with this? The current concern over the possibility of transmission of BSE from cows to humans (in the form of the syndrome CJD) may provide us with an example. Superficially, the case seems to conform to Beck's characterisation of the new large scale hazards. The topic is subject to heated and unresolved scientific controversy, and since the situation is unprecedented, the extent of the risk is literally incalculable. Given the pervasiveness of beef derivatives in the processing of many other foods, medicines and other products, and the global character of contemporary food distribution, the incidence of risk transcends spatial and social boundaries. 'Organised irresponsibility' is evident in the impossibility of tracing the source of the infection in any particular case of the new form of CJD.

However the actual course of the politics of BSE/CJD reveals something rather different from the 'risk society' account. First, the hazard was not generated by a technological advance, but rather by changes in animal feed regimes. These were adopted by the agricultural industry in pursuit of lower costs—and therefore higher profit. They were made possible by specific deregulation measures, reducing feed processing standards, introduced by the Conservative Government as part of its neo-liberal agenda. The situation was one in which a hazard already foreseen (by the 1979 Royal Commission on Environmental Pollution) was engendered by the cost-cutting *non-use* of available technology.

Second, the BSE crisis did not see the emergence of a universal interest. In the UK, the issues shifted in the space of a single day from a public health problem to a question of the protection of the interests of the UK beef industry. The response of the European Union in banning British beef exports provided an opportunity for the government, supported by large sections of the press, to redefine the issue in terms of the UK's relations with its European partners. Far from a common interest being revealed between European consumers, it became British citizens' patriotic duty to eat British beef in defiance of malevolent German attempts to damage the British livestock industry. The Labour Party acquiesced in this, confining itself to uttering concerns about job losses in the beef industry and demanding a still tougher line with Europe.

Clearly all of this illustrates the extent to which there can be no 'reading off' of perceptions of risk and responsibility from some supposedly objective 'facts'. Competing interests and discursive frames offer widely differing and conflicting 'codes' for making sense of the episode. In particular, this suggests that the focus on scientific and technological innovation, as such, as primary causes of environmental hazards, is much too narrow. Any adequate account of the BSE episode would have to recognise it as the outcome of economic, political and cultural processes interacting with one another. It would highlight the significance of the representation of farming interests in the Ministry of Agriculture, the dual role of the ministry in its responsibility for both food production and food safety, the political ideology of the Conservative government, the continuing salience of 'national' identity, and the relation between government regulation and technical advisory bodies.

Power relations which operate between and across the 'institutional dimensions', specific institutional structures, and identifiable sources of pressure and political decision, all played their part in the genesis of this particular hazard. The effect of the reflexive modernisers' abstract separation of institutional domains, together with their view of the new order of industrial hazards, is to undermine the possibility of the sort of complex empirical analysis which is needed to gain an understanding of problems like BSE. Moreover (and quite counter to the intentions of Beck, at least) the notion of 'reflexive modernisation' leads us to see episodes such as this as just further illustration of the 'inexorable' advance of 'modernity', of an impersonal process of technological development making our lives more risky. The issues are de-politicised and, paradoxically, 'organised irresponsibility' is implicitly endorsed.

More generally, where technological innovation is implicated in the genesis of ecological hazards, as in such cases as nuclear power and biotechnology, the reflexive modernisers' tendency to represent scientific and technical innovation as occurring in their own, autonomous institutional 'dimension' cuts them off from important insights available from work in the sociology of science and technology. Ever since the work of the late T. S. Kuhn, in the early 1960s, sociologists have been studying the ways in which social processes within the 'scientific community', external interests and wider cultural resources can all affect not just the rate of scientific innovation, but also its very content and direction.[8]

Similar considerations apply to technological innovation. So, if we take the case of nuclear power, the interconnections between the development of nuclear technology, economic interests, state policy in the field of energy generation, the ideology of 'modernism' itself, and military strategy, could hardly be denied: the reflexive modernisers' segregation of the political/administrative, economic, scientific/technical and military institutional complexes from one another rules out the kind of integrated analysis which the case requires. In the area of biotechnology, to take another example, there have been two notable trends in recent decades. (These are common to many fields of scientific research.) One is that overall investment in scientific and technical research has shifted dramatically away from the public sector and is now concentrated in the R&D departments of the big corporations. The second is that publicly funded scientific research is now subjected to criteria of evaluation which give high priority to anticipated commercial use.[9] Under such circumstances, to represent scientific and technical innovation as if were an autonomous process, a mere correlate of a certain phase of 'modernity', is little short of ideological mystification. In fact, the subordination of science in key sectors to the competitive priorities of private capital is all but complete.

This has an important consequence for our understanding of environmental politics. A sociologically informed understanding of the process of scientific and technical innovation renders imaginable a qualitatively different institutionalisation of science and technology. Opposition to current directions of scientific and technical change need not take the form of a backward looking, nostalgic desire for reversion to an earlier stage along a single-line developmental process, as Giddens and Beck represent it. On the contrary, the perspective suggested here emphasises the extent to which the direction of change in science and technology is currently shaped by the requirements of capital accumulation and state strategies in relation to military priorities, surveillance, control over labour processes, and product-innovation. On such a perspective, it is not required that we oppose either scientific innovation or technological invention as such. The key questions become, instead, how to detach research from its current embedding within the institutional nexus of capital and the state in such a way as to open up priorities in funding to a wider public debate, and to democratise decisions about the development and deployment of new technologies. Of course, both Beck and Giddens also

favour opening up these areas of decision-making to democratic accountability. However, their treatment of science and technology as autonomous
vis-à-vis capital and the political/administrative system sidesteps difficult
questions about the intensity of likely resistance on the part of both capital
and the state to any such project, and the immense power vested in these
institutional complexes. Only very powerful and broadly based coalitions of
social movements could have any hope of making headway with these ideas.
I shall return to this in the final section.

Environmental Distribution and the 'Death of Class'

If a recognition of the dynamics of capitalism helps to explain the particular
forms taken by science and technology, so it also casts a different light on the
proclaimed 'death of class'. The reflexive modernisers argue that traditional
class divisions and identities are no longer relevant in the face of the new
industrial hazards. Two different arguments are made here. On the one hand,
Giddens detaches environmental issues from questions of distribution by
treating the environmental movement as a response to technological and
industrial development. His position is consonant with the influential claim
made by Inglehart that environmental concern is a reflection of 'post-
materialist' values. Such values become increasingly important, it is argued,
as societies and groups within them become more affluent. The result is that
the content of the political agenda moves away from traditional questions of
distributive justice and public provision of welfare and security, towards post-
material issues such as environmental protection.

On the other hand, Beck argues that environmental issues *are* about
distribution. But in contrast to the period of simple modernity, when politics
was concerned with the distribution of 'goods' (the wealth generated by
economic growth), now the issue concerns the distribution of 'bads', particularly environmental costs of various kinds. Whereas the distribution of
goods was a battle between economic and social classes to take a larger share
of the national product (in the form of wages, profits and government welfare
spending), the distribution of bads does not follow these lines. New (and
often temporary) cleavages are created between groups *across* class lines
according to their position in relation to particular hazards.

But neither of these arguments offers an adequate account. Far from
environmental politics being part of a 'post-materialist' agenda, it can be
argued that much environmental concern is about the most basic conditions
for survival itself. Concern about the poisoning of food and water supplies,
about the quality of the air we breathe, about the danger of industrial
accidents, about the unpredictable alteration of global climates and ecosystems—these could hardly be more 'materialist'. Moreover, it is simply
wrong to argue that environmental bads do not impact on class lines. As
countless examples demonstrate, from asthma rates in the UK's inner cities
through the incidence of industrial accidents to the erosion of marginal

grazing lands in Africa, the processes of environmental degradation almost always impact most devastatingly on the poorest and least powerful communities, both within countries and globally. The rich and powerful are often able to escape the worst effects of environmental loss, whether it is through buying houses in leafy suburbs or being able to purchase raw materials such as timber from new sources when previous supplies are depleted.

Beck claims that the pattern of distribution of bads implies a qualitative break from the politics of the distribution of goods. But this again is a misleading picture of the past. Many thousands of socialist activists in their local communities and in their trade unions have been concerned with environmental health provision, with campaigning against air and water pollution, and with health and safety standards in the work-place. Engels's study of the *Conditions of the Working Class in England*, was, after all, a pioneering work of environmental socialism.[11] Socialist analysis has always emphasised the parallels between lack of 'goods', and a plentiful supply of 'bads', endemic to capitalism.

Indeed, more can be said about the relationship between specifically capitalist forms of economic organisation and ecological degradation. Economic growth (albeit uneven in both space and time) is necessary to and inseparable from capitalism. But this does not directly imply ecological destruction. In principle, changes in the composition of capital and technical innovation could allow for economic growth measured in value-terms without concomitant growth in ecological damage. However, the dominant forms of economic calculation under capitalism are abstract and monetary, subordinating to their logic substantive and 'concrete' considerations about the management of the people, places and materials involved in actual production processes. This renders capital accumulation particularly liable to unforeseen and unintended consequences at this concrete level—notably in the form of environmental dislocations of one kind or another.[12] Considerations such as these have led the Ecological Marxist James O'Connor to postulate a 'second contradition' of capitalism, to complement the 'first contraction' between capital and labour as identified by Marx.[13] This second contradiction is between the 'forces' (including technologies) of production and the 'conditions' of production, which include ecological conditions as well as human-provided infrastructures and social institutions. In short, capitalism tends to undermine its own ecological (and other) conditions of existence. In O'Connor's view, therefore, the labour and the environmental movements are both forms of response to basic structural contradictions of capitalism.

As the processes of environmental degradation under capitalism intensify through these processes, we would expect to see a wider resonance across society of current ecological concerns. But this environmental concern is increasingly differentiated in its expressions, and clear links can be seen between different definitions and policy agendas, on the one hand, and the interests of social groups and classes, on the other. In other words, the content and direction of the contemporary phase of environmental politics can increasingly be seen as an emergent arena of class conflict—but one which transforms and extends our understanding of class as it develops. Both Beck and Giddens counterpose the politics of the environment to those of class. My argument is that class politics has always, and quite centrally, especially at 'grass roots' level, been about environmental questions, and that the new agenda of environmental politics both extends and is intelligibly continuous with that longer history.

However, there is also something which transcends class politics in the new agenda of environmental politics. Beck's identification of a new order of risk does start to capture this. However, Beck's optimistic expectation of a recognised universal interest in addressing these hazards is hard to sustain. Scientific communities are increasingly aligned with interest groups in ways which make the identification and measurement of hazards permanently contested, and rival interests are affected in different ways by different policy prescriptions. Arguably the risks of 'simple' modernity retain more continuing salience than Beck acknowledges, whilst the new large-scale risks are more contentious in ways which broadly follow class cleavages than he is prepared to allow. However, awareness of the new order of risk, including the potential jeopardisation of all life on earth, has, arguably, played a part in a widespread cultural shift in recent decades. A deepening anxiety and moral horror at the scale of ecological destruction is now quite widespread. Social movements which mobilise on this basis have undergone dramatic increases in membership and mobilising capacity since the 1960s. They, and their constituency, represent yet a further possible element in a new coalition of the left, one binding together social movements organised on the basis of class interest and ones deriving from moral perspectives. Again, in terms of strategy, if not of cultural content, there is no qualitative break between such a project and the past history of coalitions of the left.

The Environmental Movement and Radical Politics

What does all this mean for our interpretation of the green and environmental movements? One is that they cannot be confined, as Giddens tends to suggest, to the role of resisters to new industrial technologies, in abstraction from the capitalist relations under which those technologies are developed and implemented. In so far as new technologies generate environmental hazards, ecological disruption and damage to peoples' quality of life, they do so as complex, culturally mediated outcomes of state policies, the product and marketing strategies of capitalist firms, and patterns of class power. Oppositional social movements are both diverse and fluid. Empirically we can observe complex and changing interpretative resources evolving within the social movements, and emergent patterns of differentiation and re-alignment. In the UK, for example, the year-long dispute between Liverpool dockers and their employer had an environmental dimension, recognised by the strikers themselves, and was subsequently supported by a coalition of road-protesters and other direct action networks. Indeed as Detlef Jahn shows in his chapter in the present volume, the formation of Green Parties in many European countries involved coalitions between previously quite diverse groupings of socialists, anarchists, civic activists, peace movement and feminist campaigners and others.

Yet it is also true to say that building coalitions between the environmental and labour movements has not always proved easy. Sharp polarities between elements in both movements have frequently emerged, especially over trade-offs between jobs and environmental protection. It is also important to pay attention to the cultural process and value-perspectives through which greens and environmentalists experience and understand processes of environmental degradation. Concern for the natural world *for its own sake*—taking seriously the place of humanity as only one species in an immense and ever-changing web of life which we have no right to destroy—is not a motivation with which those in the labour and social justice movements have always been comfortable. Even where radical greens, as is often the case, perceive 'greed' and 'profit' as the enemies of the environment, they by no means necessarily share other elements of a socialist understanding of capitalism, and generally do not see transition to a socialist society as an obvious solution (in this they are greatly assisted by the actual environmental record of the former state centralist regimes). It remains an open question whether existing patterns of *ad hoc* coalition and dialogue between greens, labour movement activists, feminists, animal rights campaigners, road protesters and so on will generate a more organically integrated and coherent new social force on the green left.[14]

Biographical Note

Ted Benton is Professor of Sociology at the University of Essex and author of numerous publications on social philosophy, history of biology, and, in recent years, on ecological issues. His recent publications include *Natural Relations: Ecology, Animal Rights and Social Justice* (Verso, 1993), *Social Theory and the Global Environment* (edited, with M. Redclift, Routledge, 1994) and *The Greening of Marxism* (edited, Guilford, 1996). He is a member of the Red-Green Study Group.

Notes

1 Anthony Giddens, *The Consequences of Modernity*, Cambridge, Polity, 1991.
2 Anthony Giddens, *Beyond Left and Right*, Cambridge, Polity, 1994.
3 *Ibid.*
4 Ulrich Beck, *Risk Society*, London, Sage, 1992; and see his chapter in this volume.
5 D. J. Lee and B. S. Turner, *Conflicts About Class*, London, Longman, 1996.
6 Ulrich Beck, *Ecological Politics in an Age of Risk*, Cambridge, Polity, 1995.
7 *Ibid.*, p. 65.

8 This is not to argue, as do some self-styled 'social constructionists', that scientifically authenticated knowledge-claims are merely cultural or linguistic constructs arbitrarily related to the world they describe. It remains possible to acknowledge the place of evidence and scientific reasoning in the shaping of scientific research agendas and knowledge-claims, while still insisting that extraneous social and cultural influences also play a significant part.

9 A. Webster, *Science, Technology and Society*, Basingstoke, Macmillan, 1991; P. Wheale and R. McNally, *Genetic Engineering: Catastrophe or Cornucopia?*, Hemel Hempstead, Harvester Wheatsheaf, 1988.

10 Ronald Inglehart, *The Silent Revolution*, Princeton, NJ, Princeton University Press, 1977.

11 Ted Benton, 'Engles and the Politics of Nature' in C. J. Arthur, ed., *Engels Today: A Centenary Appreciation*, Basingstoke, Macmillan, 1996.

12 Ted Benton, 'Marxism and Natural Limits', *New Left Review* 178, 1989, pp. 51–89; Ted Benton, 'Ecology, Socialism and the Mastery of Nature', *New Left Review* 194, 1992, pp. 55–74.

13 James O'Connor, 'The Second Contradiction of Capitalism', in Ted Benton, ed., *The Greening of Marxism*, New York, Guilford, 1996.

14 Red-Green Study Group, *What on Earth Is To Be Done?*, Manchester, RGSG, 1995.

The Quality of Life: Social Goods and the Politics of Consumption

MICHAEL JACOBS

FOR all its urge to political pragmatism in recent years, environmentalism has retained one irreducibly radical claim. This is that household consumption in industrial societies like the UK must decline. Though some may wish to play it down for tactical reasons, few environmentalists if pressed will deny their belief that present patterns of consumption are unsustainable, and will have to fall under any serious environmental programme.

It would be difficult to formulate a proposition further from the political mainstream. Personal consumption has become a dominant, even a defining feature of contemporary society, and its promotion probably the single most important objective of modern politics, more or less unquestioned right across the political spectrum.

At one time, such a radical disjuncture between the environmental programme and mainstream politics would not have bothered many environmentalists. There are still Greens for whom this is true. But in the last few years a new approach has been adopted by the more pragmatic wing of the environmental movement, aimed at shifting the ground of the environment-consumption conflict. The approach centres on the definition and promotion of the concept of 'quality of life'. In this chapter, I want to explore these debates, looking in particular at the coherence and political implications of the 'quality of life' argument.

Consumption and the Environment

The basic environmentalist position on consumption is straightforward. This is that the material and energy content of current consumption patterns in industrialised countries like the UK is too high. It is the demand for energy by consumers in the rich nations, for example, which ultimately causes the greenhouse effect: not simply direct energy use in the home and for travel, but through the energy embodied in the myriad goods and services—transported from all over the world—which make up household spending. Similarly, it is consumers' desire for cheap food which generates the water pollution, loss of biodiversity and pesticide hazards which characterise modern agriculture. It is ultimately to meet ever-rising consumer demand that forests are destroyed for paper and timber, fish stocks are being depleted and toxic chemicals released into the environment.

Of course, the rising populations of developing countries in the South of the

Published by Blackwell Publishers, 108 Cowley Road, Oxford OX4 1JF, UK and 350 Main Street, Malden, MA 02148, USA

world are also responsible for environmental damage. But environmentalists argue that this only adds to the case for consumption in the industrialised nations to be reduced. If the world's resources are limited, fairness demands that presently poor countries have priority over the already rich. And since material consumption patterns in the South follow those of the North, only if the rich countries lead by example can there be any hope of limiting the environmental damaged caused by worldwide industrialisation. Present Northern lifestyles are simply not replicable on a global basis: either the global environment, or global consumption, will have to give.

The required reductions are not small, either. Basing their calculations on the twin principles that global resource use must be held at environmentally sustainable levels, and that every citizen of the world has the right to an equal share, Friends of the Earth Europe argue that reductions of the order of 80–90 per cent are required in industrialised countries' consumption of many key resources and polluting products. Theoretical calculations relating environmental impact to consumption, population growth and resource efficiency result in similar figures.[1]

Now this does not mean that the *value* of household consumption has to be reduced by these sorts of amounts. Most environmentalists today emphasise the inefficiency of present patterns of global resource use. They point (almost as optimistically as their opponents did twenty years ago) to the possibilities of various kinds of technological and organisational advance. A combination of efficiency measures (energy conservation, waste minimisation and recycling, so-called 'clean' or low-pollution production systems, better public transport), the substitution of benign materials and activities for damaging ones (solar energy for fossil fuels, for example) and changes in the composition of final demand (labour-intensive services for material-intensive manufactured goods) will enable society to achieve a significant reduction in environmental degradation even while maintaining high levels of GDP.

But few environmentalists think that such advances can avert the need for consumption to fall at least somewhat. In practice the principal mechanism for this will almost certainly turn out to be price rises. Though environmental policy will encourage producers of goods and services to become more resource efficient and to develop less damaging products, they will be able to do this only at higher cost, which will inevitably be passed on in the form of higher prices. Some products will almost certainly be subject directly to higher taxation, energy being the most obvious example. Personal travel will therefore cost more, forcing people to do less of it, particularly by private car and air. Imported goods which have travelled long distances will likewise become more expensive. Less chemical-intensive agricultural production methods will raise food prices. More strictly regulated production regimes will raise the prices of timber and paper. And so on. Under a strong environmental programme, therefore, consumers may find themselves earning as much as before, but their real income, measured in terms of purchasing power, will have declined.

It is possible that in this process the prices of some goods and services will fall. If one instrument of environmental policy is the so-called 'eco-tax reform' (see Stephen Tindale's chapter in this volume), price rises for highly-taxed energy-intensive goods may be compensated for by price reductions for labour-intensive services on which taxation has been cut. But it is not clear how consumers will react to this. Effectively, environmental policy will have forced them to change their consumption patterns. They may be able to go out to the restaurant more, but they won't be able to go on holiday so often. Most environmentalists acknowledge that, whatever people's income levels, such enforced changes in *lifestyle* may well be perceived as a reduction in living standards. They have no illusions that the transition to a sustainable society will be painless.

Consumption Politics

This prospect sets environmentalism squarely against the forces of modern politics. The role of consumption in political life can hardly be under-estimated.

Today raising the level of consumer spending is generally regarded as the key political objective for any government: the principle measure of its success and—according to the conventional wisdom of psephological predic-tion—the best indicator of its likely vote. Recent political history has been widely understood in these terms. It was the expectancy of rising consumer spending in 1992 which (it is generally believed) sustained the Tory vote against the odds in that year's general election; its decline in the mid-1990s the cause of the Conservatives' subsequent slide in the opinion polls. The relationship between the two indeed gave this period one of its most persistent political themes. As the resumption of economic growth failed to bring with it the return of consumer confidence, politicians and media commentators began the search for the missing 'feelgood factor' which would apparently determine the outcome of the general election. When consumption levels finally rose at the end of 1996 (though to the puzzlement of the commentators, without much associated 'feeling good') the sigh of political relief was almost audible. 'As promised . . .', ran the Tory pre-election poster, '. . . more spending money'.

And it is the desire to increase consumer spending which surely lies behind the obsession with income tax rates which now characterises British political life. In the 1980s successive Conservative governments justified the reduction in income tax on the grounds that households should have more of 'their own' money to spend. Although dubious as a fiscal strategy, the political success of the argument can hardly be doubted. After the 1992 general election, when Labour's small proposed income tax rises became the centrepiece of the Tories' successful assault against them, Labour was forced onto the Con-servatives' territory, unwilling to oppose income tax reductions. The 1997 election saw both the major political parties parading income tax-cutting plans, competing to claim that they would reduce income tax the most.

The reason is surely obvious. Taxation takes money from consumption, and consumption is now the highest political good.

The Green Case

It is into this context then that the environmental argument is propelled. The circumstances, it would seem, could hardly be less propitious.

For more radical Greens, this is not a problem. Unconcerned to win mainstream political support, they are happy to challenge the whole culture of consumerism head-on. They insist not only that Western consumption patterns are environmentally unsustainable, but that they are anyway undesirable. Driven by capitalism's imperative for continuously expanding demand rather than any relation to the meeting of human needs, consumption growth does not make people happier. On the contrary, seductive advertising and social pressure lead consumers into a spiral of competitive purchasing, to pay for which they are forced to work more and longer hours, thereby reducing the time available to enjoy life for itself, and increasing stress and ill-health. People would actually be better off, Greens argue, if they consumed less and concentrated more on genuine wellbeing: on personal development, on relationships with others and on social belonging. A new term—'downshifting'—has even been coined for those who take this path.[2]

The environmental evidence of excessive resource use then becomes a moral injunction to individuals to consume less. Radical Greens stress each person's responsibility for their own contribution to globally unsustainable consumption. Hence the familiar lifestyle choices urged by books on 'how to be Green'. Save energy; recycle waste; repair, don't replace; cycle, don't drive; eat local, seasonal foods not exotic imports; buy durable goods, second-hand where possible. 'Live simply', as a favourite Green slogan puts it, 'that others may simply live'.

As moral principle (and social example-setting) this is admirable; but as politics it is surely hopeless. Indeed, the radical Green position is barely a political argument at all, in the sense of a claim addressed to society's collective agency. It is a personal appeal, addressed to individuals and their own sense of individual wellbeing. And as such it is in a strong sense misdirected.

For one thing, it is almost certainly not true that consuming less will make people feel better off. For the individual on any income below the very highest, having more money to spend increases the range of opportunities available for personal development as much as for 'mere' material consumption. To deny that leisure activities—sports, music, books, travel and tourism—constitute genuine means of meeting higher human needs is surely to take an excessively austere view of human life and its pleasures. And it is increasingly on these goods that higher spending goes.

50

It is true that much of it also goes into what might be called 'competitive consumption', the purchase of ever-higher-quality goods and of fashionable items whose value lies almost entirely in the desire to 'keep up with the Joneses'. I have to buy a new car, because everyone else has done so; not to do so would be to lose social status even if its 'objective' additional value is minimal. Anyone with a child who has demanded a new pair of trainers will recognise the pervasiveness of this phenomenon of competitive social comparison. But to deal with this is surely useless to ask individuals to reduce their consumption voluntarily or to see themselves as better off if they do not play the comparative game. This is to view people in isolation from the society to which they belong; or to ask them to break from this society to join another, minority one of low-consumers.

As studies of 'downshifting' show, very few people are able to do this: the pressures of social conformity (indeed, of 'belonging') are too great. The mistake here is to adopt the neo-liberal assumption that the consumer is a sovereign individual, making autonomous choices in free markets. In fact, contrary to appearances, consumption is not a purely individual form of behaviour. It is a social force in which individuals are bound up, and only those with strong will and alternative social networks are able to escape. Appealing therefore to people as individual consumers to reduce their consumption (or to allow it to be reduced) on the grounds that having less will make them individually better off is likely to be futile for all but a few. It seeks to make shoppers into buddhists, surely an alchemic transformation.

In fact, environmentally sustainable consumption will not come about through individual choices, but through regulatory policies collectively decided and imposed by the state. To support these the public must be appealed to, not as consumers, but as voters. Admirable as voluntary reductions in consumption are, they are not the route to environmental improvement.

The Quality of Life

This is why in recent years pragmatic environmentalists have increasingly used a different kind of argument to express their concerns about the environmental effects of consumption. They have focused, not on what consumers will *lose* as a result of environmental policies, but on what they will gain. Such policies improve the environment: they lead to lower greenhouse emissions, reductions in pollution and traffic congestion, protection of countryside and natural habitats, and so on. The argument then is that these things *make people better off*.

Of course, they do not do so in the conventional sense. They do not increase the 'standard of living' as measured by personal disposable income. They are not components of private consumption. But this, the environmentalist argues, only exposes the inadequacy of these concepts as measurements or definitions of personal wellbeing. It is not only private consumption which makes people well off, but also a range of other goods which individuals enjoy but do not personally buy. Environmental goods—clean air, low traffic levels, protected countryside—fall into this category. If, to pay for such goods, prices and taxes must rise and private consumption must therefore be somewhat reduced, this does not automatically mean that people are worse off. They may have less money to spend, but their overall *quality of life* may be higher. Improvements to environmental goods may outweigh the marginal reductions in private consumption required to pay for them.

In this way the concept of 'quality of life' plays a crucial role in the new environmentalism. It challenges the politics of consumption side-on. It does not attempt to deny that consumption contributes to wellbeing. It claims merely that other goods also contribute; and that the goal of public policy should be to raise the overall level. Environmentalists of course add an implicit additional argument: that trading off private consumption for environmental goods not only *may* but *would* in fact increase overall wellbeing.

The concept of quality of life encompasses more than just the environment. The environment is an example (or a collection of many different examples) of what we may call *social goods*. Social goods have two key characteristics. First, they are shared. The air I breathe is the same air you breathe; the landscape is our common heritage. Secondly, and because of this, social goods must be provided collectively. However rich I am, I cannot buy clean air with my own income; only if I am exceptionally rich can I buy the landscape. If we want these things we must secure them collectively. In most cases this must be done either through taxation and public expenditure or through economic regulation. This is true of nearly all environmental goods.

Environmental goods are paradigm cases of social goods; but there are others, equally or more politically salient. Personal safety—freedom from crime and from the fear of crime—is a social good. (Or another way of putting this is that crime is a social cost. Social costs and social goods are opposite sides of the same coin.) It is shared amongst the members of a society. If crime levels are low, everyone in the neighbourhood feels safer. And it can only be secured collectively. Individuals cannot purchase personal safety with their own private income. Indeed, attempts to do so—locks and anti-burglar devices, personal security alarms, and so on—frequently make people feel less safe, not more. Real personal safety can only be achieved through collective action: particularly through policing (in various forms) but also, it seems clear, through the reduction in equality which generates the conditions for crime. Such collective action, in most cases, requires public expenditure. That is, if people want to feel safe on the streets and in their homes—and the evidence suggests that in recent years people have come to feel increasingly unsafe, and that this has reduced their sense of personal wellbeing—they must expect to pay for it through taxation.[3]

Of course, public services and facilities in general are also in this way social goods: the education and health services, public parks, libraries, street cleaning, subsidised theatres, and so on. These too are shared and provided collectively. They too rely on taxation. The importance of the concept of quality of life is that it places all these social goods—both public services, and the less tangible goods of environmental quality and social order—in direct relation to private consumption. It makes a simple claim: that social goods also contribute to overall wellbeing, and that over-emphasis on private consumption at the expense of social goods may actually lead to a decline in wellbeing rather than to its increase.

Arguably, this claim offers an alternative framework for understanding certain trends in British society and politics in recent years. Average post-tax disposable income and private consumption have unquestionably risen since 1981. But this has been achieved, it can be argued, at the expense of declining social goods (or rising social costs). In the case of public services, the relationship has been fairly explicit, tax cuts having taken obvious precedence over public spending. The sense that the National Health Service and the education system are in crisis is widely felt. But the less tangible social goods of environmental quality and social order have also declined. Air pollution and traffic congestion have become worse. Crime and the fear of crime have risen. Indeed other, even less obvious social goods may also enter the equation: opinion surveys show that people have become more anxious about the future, and less happy with the condition of British society as a whole; its fairness, the functioning of its democracy, its morality.[4]

In this sense, the argument runs, the overall *quality of life* in Britain has declined even while the private 'standard of living' has increased. This might be held to explain the events of 1996–7: increases in private consumer spending failed to restore the so-called 'feelgood factor' because private consumption is not sufficient to make people 'feel good'. By 1996 the decline in social goods had come to outweigh the rise in private income, leading to a reduction in the overall quality of life. Though most people would not have articulated it in this way, this reduction was what they experienced. The Conservative government was blamed.

This kind of argument would seem to have particular political resonance in cities. It is in urban areas that the decline of social goods and the increase in social costs are felt most starkly. Most people living in cities are only too aware of air pollution, and the risks to health it poses. (One in seven children is now reported to suffer from asthmatic symptoms.[5]) They endure traffic congestion daily. Crime and the fear of crime are highest in cities: men as well as women report in survey research that it affects their everyday behaviour.[6] It is in urban areas that public spaces appear dirtiest and least welcoming; that public services feel most in crisis. In these senses the negative effects of social costs on individuals' wellbeing is almost certainly more starkly felt by people living in cities than by those in smaller settlements and rural areas. The concept of quality of life would seem to be particularly useful here, giving political articulation to common experience.

The Politics of Taxation

An important proposition follows from these arguments. It is that taxation can make people better off. If taxation is used to pay for social goods, this can improve the overall quality of life, even though it demands a sacrifice in private consumption. In this sense the concept of quality of life offers a direct challenge to the conventional political wisdom which asserts that, by taking money from private consumption, taxation necessarily makes people worse off, and should therefore wherever possible be minimised. As social costs rise, the marginal contribution to wellbeing made by public spending on social goods (which individuals cannot purchase privately) can come to exceed the contribution made by private spending. This is the condition, the environmentalist seeks to argue, in which Britain and other industrialised countries now find themselves.

Is it in fact true to say that social goods such as cleaner air and lower crime cannot be purchased privately? Within particular neighbourhoods this statement would appear to hold; these goods are shared. But social goods are not equally distributed between neighbourhoods. Air quality, crime, traffic congestion and public services and facilities vary between different areas. For households which are mobile, this does make them effectively purchasable. If you want better air quality, you can move to an area which has it. The premium on house prices in small towns and suburbs with better air quality and lower crime (and even more with better state schools) reflects the fact that many people are doing just this.

But it also acts to limit this possibility. As house prices in areas of low social costs rise, the proportion of the population able to afford them declines. (The availability of employment means that only a minority of households are in any case mobile.) This effect is exacerbated as the prevalence and severity of social costs increase. There are fewer and fewer areas which do not suffer from poor air quality, traffic congestion and crime. Small towns and suburbs are certainly no longer immune. Public services are under pressure almost everywhere. As the decline in social goods becomes more widespread, so the possibility of buying one's way out of them recedes. (Where it remains possible, such goods are often purchasable only at the expense of other social costs, such as longer commuting distances.) Increasingly, therefore, this will become an option only for the very rich. Such social goods will have become classic cases of 'positional goods': increasingly scarce, highly priced, and available only to a minority.[7]

This prospect would appear to strengthen the argument for taxation to secure the public provision of social goods. In recent years it has been widely argued that the popular basis for high levels of public spending has been more or less permanently destroyed. According to J. K. Galbraith and others, the post-war social democratic settlement has been undermined by its own success. Public spending now benefits mainly the poor, so the 'contented' majority are no longer prepared to support it through taxation.[8] But the increasingly widespread experience of environmental and other social costs appears to contradict this argument. If a majority of people, including those on average incomes, are now experiencing such costs, and are increasingly unable to buy their way out of them privately, it may be possible to recreate a political basis for public spending after all.

It is notable that the concept of quality of life links environmentalism to other political causes. It provides a general rationale for public spending and for regulatory economic policy. Indeed, environmental protection here becomes just a subset (though a very important one) of the wider class of policies aimed at securing social goods—whether better policing and measures to reduce social inequality (to reduce crime), or various kinds of expenditure on public services, or political reforms aimed at restoring trust in the democratic process. It does not *require* that environmentalists place themselves in alliance with those desiring other social goods; but it provides a common framework of argument to unite these groups if they so wish. This is indeed exactly what appears to be happening under the 'Local Agenda 21' initiative at local level, where environmental concern is being tied to a wider set of social goods (including health issues and poverty reduction) under the general label of improving quality of life.

The Democratic Deficit

In these ways, the concept of quality of life appears to give modern environmentalism a new way of arguing: one which reflects the experience of rising social costs in recent years, which challenges squarely the priority given to private consumption in contemporary politics, and which offers the opportunity to build wider political alliances.

Unfortunately, things are not quite as simple as this. For two reasons, this straightforward version of the 'quality of life argument' is not a sufficient basis for a successful environmental (or wider) politics.

The first reason is that the link between the existence of social costs and the desirability of taxation (or environmentally-induced price rises) is highly contingent. Let us leave aside the possibility—which is surely a very real one—that many people would prefer (say) to suffer current air pollution and traffic congestion levels than be forced to reduce their car use by tax or price rises. Even if voters accepted such an exchange in theory, they cannot guarantee that it will occur in practice. A government may claim that the effect of higher taxes or prices or other traffic measures will be to reduce air pollution and congestion. But to win public support voters must accept this in advance of it actually happening. For good reasons they may be sceptical.

For one thing they may doubt whether there is in reality sufficient knowledge about the relationship between the environmental problem and the available policy measures. Will transport behaviour in fact change as predicted if taxes and prices rise? If not, citizens may find themselves paying the extra *and* suffering the pollution and congestion. Do we know whether higher energy prices will reduce carbon dioxide emissions enough to slow global warming? The answer depends on many factors outside the government's control. Given scientific uncertainty and previous policy failures, scepticism about the ability of governments to influence environmental problems is surely understandable. The same is perhaps even more true in the social field. However, much is spent on improving policing and attacking poverty, there is no certainty that crime will be reduced.

Even more importantly, voters must take it on trust that governments will actually do what they say they will do; that they will use the money raised from higher taxes to deal with the environmental and social problems promised. Here the political argument runs into an even more profound disillusionment. Public attitude surveys show that trust in the institutions of government in Britain has declined markedly.[9] If people do not trust governments to pursue their promised objectives, they are unlikely to accept the quality of life argument. The argument effectively offers a gamble: exchanging the certainty of reduced private consumption for the uncertain benefits of public expenditure and regulation. It is a gamble many voters may reasonably reject.

For this reason the quality of life argument cannot stand on its own. To be politically effective it must almost certainly be related to some kind of parallel argument about the need for a renewal of trust in government itself. Here too British environmentalism will find itself in natural alliance with an apparently separate political cause.

This is not the place to rehearse the wider issues of democratic reform. But two specific suggestions can be made. One is for the earmarking or 'hypothecation' of taxes. If one of the causes of public disillusionment is the sense that taxes are collected but voters do not know or cannot see how they are spent, the proposal that some taxes or proportion of taxes should be earmarked for specific purposes would appear to have some merit. It would reconnect taxation to the purposes for which it is raised, and would therefore increase the transparency of the political transaction.[10] Partial hypothecation of environmental taxes for public transport, energy conservation and pollution reduction is widely canvassed; but earmarking a certain proportion of taxes for crime reduction, health and education spending and other social goods would also surely add to their political legitimacy.

The other suggestion concerns the use of indicators and targets. The political salience of the quality of life argument rests on the clarity of the relationship between social goods and personal wellbeing. It is only if people can perceive how social goods have declined that they become conscious of the effect on their quality of life. Developing appropriate sets of indicators through which trends in social goods can be measured and publicised is therefore likely to be particularly important. By using these indicators to set targets for the improvement in social goods, establishing the link between particular policies and their effects on different aspects of quality of life, governments may find it easier to win public acceptance. The development of quality of life indicators is indeed now proceeding apace in local government and NGO circles.[11]

Individual and Social Wellbeing

Even if the problem of the 'democratic deficit' can be addressed, however, this does not give the simple quality of life argument a free run. The second reason for doubt goes right to the heart of the concept.

If we examine the social goods which contribute to the quality of life—particularly the environmental goods—it is not clear that all of them can be said to contribute to *personal* wellbeing. Personal wellbeing is the satisfaction one gets from feeling that one's own life is going well: that one is healthy, comfortable, secure, purposeful, autonomous, that one has good relationships with family and friends, and so on. Private consumption generally contributes to personal wellbeing: it is to achieve (often small) improvements to these qualities that we purchase goods and services for ourselves. (We may of course be mistaken about the contribution they will make: various theories of human needs suggest that we often are.)

Now some social goods clearly also make contributions to personal wellbeing. People who experience crime or the fear of crime are almost certain to feel less secure and that their autonomy is restricted. If I have children, the quality of the education system affects my wellbeing in a direct and personal way. Some environmental goods are of this kind too. The quality of the air I breathe concerns me very personally: my health, and that of my children, is directly affected by it. Similarly, if a stretch of countryside I know well and often walk in is spoiled or destroyed by development, or an often-used beach ruined by sewage pollution, this may affect my personal wellbeing at least as much as, if not more than, any consumer good or service I might buy.

But this is not true of all social goods. If I do *not* have children, the quality of the education system does not affect me personally, and cannot be said to contribute directly to my personal wellbeing. If I am not a theatre-goer, or a user of public libraries, the same will be true of changes to these services. The quality of rivers and natural habitats, and of parts of the countryside I do not know and do not visit, similarly do not contribute to my *personal* wellbeing. To say that *I* am made better off by these things is surely incorrect. *My* wellbeing is not enhanced by the continued existence of rare plant species on the other side of the world.

And yet most of us do not want simply to dismiss this kind of statement. There is clearly something in the idea that we are made better off by these things, by living in a society with a good education system, theatres and libraries, and in a world which contains a wide diversity of life. But that is the point: it is 'we' who are made better off; it is not 'I'. Many social goods do not make us better off as individuals; rather, they help to create a better society in which we as individuals then live.

57

This is surely, a crucial distinction. In recent years the dominance of a neo-liberal economic philosophy has helped to establish the belief—rarely articulated as such but implicit in much political discourse—that the wealth and wellbeing of a society consists simply in the sum of the wealths and wellbeings of its individual members. If, as Mrs Thatcher famously put it, 'there is no such thing as society, only individuals and their families', there simply is no larger entity whose wellbeing can be independently defined. A prosperous society just is one in which its individual members are prosperous. Hence the emphasis in current political discourse on household consumption: making society better off means, straightforwardly, making its individual members better off.

Environmentalists and other defenders of social goods have a strong interest in resisting this view. They need to be able to say that there are things which make a society or a world better off (or simply better) whose value does not rest on the benefits accruing to individuals. Many environmental goods—natural habitats, species, landscapes, river and marine water quality—fall into this category. They are not directly experienced by many individuals, and they cannot be said to improve many people's *personal* wellbeing. Yet they have value: they contribute to the goodness or flourishing of society or the world as a whole. Environmental economists have called this 'existence value'. Most people, they have found, are willing to pay for environmental goods from which they gain no direct use themselves. They want rainforests and blue whales and unspoilt wilderness areas to exist, even though they will never see them or have any other use for them.

Other social goods are also surely like this. The value of academic research, museums, art galleries, libraries, public service broadcasting and subsidised arts does not lies solely in the pleasure or enlightenment they give to individuals. The number of people using them is important, no doubt; but it is not the criterion by which their value is judged. They should not be closed down when 'insufficient numbers' use them. Intellectual and cultural goods of these kinds contribute to the shared wealth of society; indeed, to its very identity. In this sense they are shared even by those who do not benefit directly from them. We all gain from living in a society which is rich in culture and learning—and in nature.

The same is true of what we might call 'moral goods': justice, equality, the eradication of homelessness or compassionate treatment of asylum seekers. For those benefiting, these things are of course private goods, contributions to personal wellbeing. But for others, they are things which make society a better place to live in. They do not contribute to my personal wellbeing; but I nevertheless want to live in a society where such goods exist.

All this has a rather important bearing on the quality of life argument. It means that environmentalists cannot simply argue that protecting social goods and reducing social costs improves personal wellbeing. Only for some issues is this true: those such as air pollution, traffic congestion and crime which people directly experience. In relation to other social goods—

including some of the things, such as natural habitats, environmentalists most want to defend—such an argument will surely have little resonance. Arguing that voters will be personally better off if they sacrifice a little private consumption to protect natural habitats (or to reduce homelessness) simply invites rejection.

What is needed, rather, is an expansion of the concept of quality of life. This must refer not simply to the quality of individuals' lives, but to that of our collective life as a society. The argument for quality of life, that is, must appeal beyond the interests of individuals in their own personal wellbeing to their interest in the wellbeing of society or the world as a whole.

These are of course related. Many people will accept that it feels better to live in a better society. In this sense, no doubt, my personal wellbeing is affected by the wellbeing of society. But they are conceptually distinct. The value of private goods lies in the benefits they give to me. But the value of a good society does not rest on the good feelings I get from living in it. If it makes me feel better, this is because I recognise that it *is* better; not the other way round.

More importantly, it isn't clear that in their political outlook most people *do* identify their own wellbeing quite this strongly with the condition of society as a whole. The promotion of neo-liberal ideology in recent years would appear to have undermined this outlook, encouraging people to concentrate on their own and their families' wellbeing, to be less concerned with the condition of the wider society. This is surely why the promotion of private consumption has become such a dominating feature of contemporary politics. But it is precisely a sense of wider social identification which the quality of life argument needs to promote. It needs people to feel that there is such a thing as society, and that its wellbeing matters to them.

In other words, the concept of quality of life seeks not just an expansion of the *objects* of wellbeing, from private goods to social goods, but of the *subject* of wellbeing too. It asks people to consider themselves not just as individuals, with private interests, but as members of society, with social interests too.

In this way the quality of life argument attaches environmentalism firmly to a wider political goal: to the reassertion of the individual as a member of a community. In a society increasingly dominated by market relations, in which individuals have been cast primarily in the role of consumers, it seeks to re-emphasise the inescapably social basis of personal identity, and therefore of political purpose. It reinterprets environmentalism, that is, as an essentially communitarian project, concerned with the value of the things we share, and the need to defend them in a reinvigorated public sphere.

Conclusion

The important question, of course, is whether all this can be made to 'work' politically. Can the argument for quality of life win the wider political support which environmentalists will require in order to address the unsustainability of present consumption trends? We do not know, of course. But it is surely not implausible to think that it might. The present author's research, using focus groups, suggests that most people do in fact recognise the importance of social goods, and give the environment particular priority. Moreover they do have a conception of themselves as members of society and not simply as autonomous individuals. What they do not have, in general, is a language in which to express these ideas *politically*. The term quality of life itself is not yet familiar in this meaning. In this sense the task for environmentalists would appear to be at least as much that of *articulating* the argument for quality of life as it is of persuading people of its merit.

Biographical Note

Michael Jacobs is a Research Fellow in the Department of Geography at the London School of Economics and Political Science. He is author of *The Green Economy: Environment, Sustainable Development and the Politics of the Future* (Pluto Press, 1991) and *The Politics of the Real World* (Earthscan, 1996).

Notes

1 Friends of the Earth Europe, *Towards Sustainable Europe*, London, Friends of the Earth, 1995; Paul Ekins and Michael Jacobs, 'Environmental Sustainability and the Growth of GDP: Conditions for Compatibility', in Andrew Glyn and V. Bhaskar eds., *The North, the South and the Environment*, London, Earthscan, 1995.

2 See Polly Ghazi and Judy Jones, *Getting a Life: The Downshifter's Guide to Happier Simpler Living*, London, Hodder & Stoughton, 1996.

3 L. Dowds and D. Arendt, 'Fear of Crime', in R. Jowell *et al.* eds., *British Social Attitudes: The 12th Report*, Aldershot, Dartmouth Publishing, 1995. It is true that in some parts of the United States, low crime rates have effectively been privately purchased by social exclusion. Shopping malls are privately policed and 'undesirable' persons kept out. Whole residential estates have now been built with walls, gates and security guards, paid for by a form of 'private taxation' of residents. But quite apart from the practical and social difficulties of creating such estates in Britain, it is not clear that this is in practice an effective strategy to reduce the fear of crime. Few people will want to live their lives solely within fortress walls; outside them, lower crime is still dependent on public policing and the reduction of inequality.

4 MORI State of the Nation Survey for the Joseph Rowntree Reform Trust, published in *British Public Opinion* (MORI Newsletter), June 1995; MORI poll conducted for Barnardo's and published in *The Facts of Life: The Changing Face of Childhood*, London, Barnardo's, 1995.

5 National Asthma Campaign, 1995.

6 Dowds and Arendt, *op. cit.*

7 Fred Hirsch, *Social Limits to Growth*, London, Routledge and Kegan Paul, 1977.

8 John Kenneth Galbraith, *The Culture of Contentment*, London, Penguin, 1992.

9 MORI State of the Nation survey, *op. cit.*; J. Curtice and R. Jowell, 'The Sceptical Electorate', in Jowell, *op. cit.*

10 See for example Geoff Mulgan and Robin Murray, *Reconnecting Taxation*, London, Demos, 1993.
11 See Alex MacGillivray and Simon Zadek, *Accounting for Change: Indicators for Sustainable Development*, London, New Economics Foundation, 1995.

Environmental Politics: The Old and the New

JONATHON PORRITT

THE environment movement (by which I mean that aggregation of individuals and disparate organisations engaged in environmental issues, broadly or narrowly defined, at the local, national or international level) is currently going through one of its periodic bouts of repositioning. It is a time of considerable intellectual ferment, shaped by a widespread but still fuzzy consensus that 'we are moving into new territory', and an equally widespread lack of consensus as to what the contours of that territory will turn out to be. Twenty-five years on from the first UN Conference on the Human Environment in Stockholm, and five years on from the Earth Summit in Rio de Janeiro, it is as good a time as any to seek to identify what it is that is genuinely new about the environment movement today, and what is either constant or simply more of the same rehashed for different times.

Taking Stock: Environmentalism Entrenched

The first thing to be said is that it is astonishing that the environment movement has weathered the last eighteen years so well. Not only has it moved out of the fringes to somewhere near the centre of the political stage, but compared with the diminution there has been in the trade union movement, or in the broad case for a socially progressive, redistributive agenda, both nationally and internationally, the environment movement has made real and sustained progress in an ideologically hostile context.

There are several possible reasons for this. The first is the very depth and diversity of the environment movement which gives full rein to the instinctive opportunism and tactical flexibility of many organisations. This flexibility is in turn possible because of the relative lack of ideological baggage which characterises the positioning of the vast majority of environmental organisations. Accusations that environmentalists are little more than 'crypto-socialists' have never really stuck in this country, and Mrs Thatcher's charge in 1986 that environmental organisations were part of the 'enemy within' was short-lived and regularly belied in practice by surprisingly cordial relationships between environmental campaigners and government Ministers and officials.

More important than either of these is the fact that what people feel about the environment, however ill-informed, ambivalent or inconsistent it may be, is very often a 'crystallisation' of much deeper feelings about the kind of society they find themselves living in. As Robin Grove-White has said, 'a

© The Political Quarterly Publishing Co. Ltd. 1997
Published by Blackwell Publishers, 108 Cowley Road, Oxford OX4 1JF, UK and 350 Main Street, Malden, MA 02148, USA

key reason the environmental issues have had such continuing resonance in countries like our own is that they have provided an umbrella and a shared vocabulary for reflecting a range of problems, anxieties and tensions that lie deep within modern industrial society.'

For all its resilience and flexibility, the environment movement is now hard up against some extremely demanding challenges. The easy bit (getting everybody to pay attention to the state of the world and to accept in theory the need to change our ways) is over. Turning that theoretical consensus into operational practice is infinitely harder. What is more, as Geoff Mulgan, the Director of Demos has often pointed out, environmentalists may have become victims of one aspect of their own relative success over the last twenty-five years: having chipped remorselessly away at the authority of politicians, business people, economists, scientists and engineers, they now find that they need the resources and reach of those very institutional forces to get working solutions delivered on a sustainable basis. Some authors in this collection take a different but even more disturbing line in suggesting that the environment movement may have lost its capacity for genuine radicalism by allowing so much of its work to be 'framed' by political and scientific orthodoxies that are now being questioned in a fundamental way by society at large.

That of course is the thinking Green's rather elegant version of what others have described more bluntly as the progressive betrayal by many environmentalists of both radical protest and radical policy alternatives. Leaving such rhetorical charges to one side, there have indeed been many compromises made, particularly since the late eighties; and part of the purpose of this chapter is to explore the degree to which these compromises can be justified. What may need to be done to offset the potentially damaging consequences of such compromises in terms of lost authority, integrity or intellectual radicalism? One thing is clear, however: this particular aspect of the modern environment movement is as old as the movement itself. In one form or another, the 'dark green' versus 'light green' or purists versus accommodationists has been part of the environmental scene since the time when Gifford Pinchot and John Muir fought it out over the fate of America's old-growth forests back in the nineteenth century. It is a debate which is becoming even more charged as the condition of the global environment deteriorates. This is obviously not the place to be carrying out any comprehensive review of the state of the earth, but it is important briefly to recall the full scope of the ecological pressures currently bearing in upon us, not least because a vocal and well-funded cabal of so-called 'contrarians' (working predominantly out of the USA, but in Europe as well) continue to assert that all is well with the world and that environmentalists are nothing more than latter-day Malthusians lining their own pockets by exploiting peoples' fears.

You have to suffer from a special kind of blindness to be able to ignore the weight of hard evidence which cumulatively reinforces the broad case that has driven the environment agenda for the last twenty-five years: that the natural systems which govern life on Earth are now subject to ever greater anthropogenic stresses, threatening serious disruption in the next century and already condemning many millions to chronic poverty and ill health.

An increasingly powerful consensus has grown up around this analysis over the last twenty-five years, embracing governments, academics and a vast array of non-governmental organisations. It is not really in dispute any longer that the way we are living now is literally unsustainable, or that we should embrace as rapidly as possible a development path that *is* sustainable in terms of the life support systems on which we depend. But as Michael Jacobs has pointed out, it is not all that difficult (knowing what we now do) to establish consensus at that abstract level; it is in the policy development and implementation that the concept of sustainable development becomes contestable rather than consensual.

That said, most environmentalists would dearly love to see a lot *more* contestation within and between the major political parties in the UK; it is their relative indifference to this agenda which is much more worrying.

Having exonerated both the Liberal Democrats and Plaid Cymru (who have made genuine and consistent efforts to embed some real sustainability thinking at the heart of their politics), commiserated with the Green Party (which remains incapable at the national level of overcoming the inequity of our first past the post electoral system), and extolled the burgeoning enthusiasm of a very large number of local authorities in the UK who have now committed themselves to turning the rhetoric of sustainable development into practical measures at the local level, there is really not a great deal more to be said about the contribution of the UK's political parties to the development of environmental ideas and practice.

With the crucial exception of Mrs Thatcher's short-lived 'green period' in the late eighties (which really did raise the stakes in a quite dramatic fashion), there has not been an ounce of heavyweight political leadership on environmental issues over the last twenty-five years. There are few who suppose that this will change even after a change of Government.

Environmentalists and Liberal Democracy

Despite widespread disillusionment throughout the environment movement with our two major parties, there is still overwhelming consensus that the best way forward lies within the UK's democratic, parliamentary procedures and systems.

The thinking (or, more often, the unexamined assumption) behind this consensus is simple. First, there are intrinsic qualities within our inherited model of democracy, however flawed and anachronistic that model may be, which confer such benefits on society as to make it unthinkable to lose them.

Secondly, given the extent of the life-style changes and the depth of the production revolution which will be necessary to ensure a peaceful transition to a genuinely sustainable society, the very idea of supposing that such a society might be imposed by authoritarian *diktat* borders on the surreal.

Thirdly, for what it is worth, an overwhelming majority of environmental activists, writers and philosophers over the past twenty-five years have been staunch defenders of the values of liberal democracy, notwithstanding the almost perverse fascination of academics and commentators with the handful of activists, writers and philosophers who have chosen to argue that even the fruits of liberal democracy must be subordinated to the achievement of a sustainable future for our species.

Perhaps they go on about it so much partly because environmental NGOs have taken so little *direct* interest in constitutional and civil rights issues. Prior to Friends of the Earth's pioneering advocacy in 1995 of the need for proportional representation—as the first of a string of necessary reforms— little comes to mind other than a string of rather tokenistic press releases about the Criminal Justice Act. The emergence of Real World (see later) has at least started to make the linkages between a healthy environment and a healthy democracy somewhat more transparent. But it is still surprisingly hard work.

After all, environmental NGOs rely to a very large extent on access to a functioning, reasonably accountable democracy. A significant proportion of the human and financial resources of the environment movement is still devoted to the mutually reinforcing activities of lobbying political parties directly or 'consciousness raising' amongst the general public to increase the pressure on political parties indirectly. It is universally accepted amongst environmental organisations that pressure of this kind means digging in for 'the long haul', in as much as pressure groups cannot afford to move too far ahead of the political parties, and the parties cannot afford (or don't dare) to move too far ahead of the electorate.

As it happens, the long haul continues to bear much riper fruit than most people realise. A comparison between the emerging transport policy of the Conservative Party in its last eighteen months in office and the campaign demands of Friends of the Earth a decade ago, makes interesting reading. A national cycling strategy; the Safe Routes to School scheme; encouragement for local authorities to shift more resources into public transport; the open acknowledgement that 'demand management' is now a crucial aspect of transport planning; all have been duly recycled ten years on.

But the long haul continues to leave many activists terminally frustrated and intent on more direct and often unlawful actions to speed the process of policy change. Again, there is nothing new in this; environment movements the world over have *always* had direct action elements of this kind, although there are some commentators who believe that there are now more direct action groups confronting environmental and social 'wrongs' than ever before.

Interestingly, there is *no* automatic consensus amongst such groups about the intrinsic benefits of representative liberal democracy. Indeed, many of the processes and outputs of that democracy are challenged head-on, as inequitable and ill-informed, sometimes without any regard whatsoever for the level of public support that may underpin the decision of elected politicians at local or national level. A 'higher moral order' is often invoked. Though there are many (including myself) who think that such a challenge is right and proper in certain circumstances, the implications of this are rarely, if ever, discussed.

Indeed, the whole debate about the contribution of direct action campaigns to the modern environment movement often seems completely detached both from the history of the movement and from reality. Those who passionately aver that it is in these 'sub-political' campaigns that the seeds of the 'new environmentalism' are to be found are no less detached from reality than those who argue that these campaigns portend the very breakdown of civil society and an inexorable slide into anarchy.

The barrage of hyperbole at the margins serves little purpose other than to keep the media interested. The reality is much more humble: that today's direct action campaigns are almost as established a part of the modern environment movement, and almost as well accepted by the public, as the mainstream NGOs. There is no evidence whatsoever that their activities erode the credibility of the rest of the environment movement, as some have claimed, by tarring all with the same brush of anarchic confrontation.

A Gallup Poll at the beginning of June 1995 revealed that 68 per cent in this country would be prepared to entertain the idea of civil disobedience in defence of a cause they believed in, 14 per cent up on a similar poll in 1984. Of Tory voters, 52 per cent agreed with the proposition that there are times 'when people may be justified in disobeying laws to protest against things they find very unjust or wrong'.

This caused *The Daily Telegraph* to take the majority of its own readers seriously to task, likening them to the Gadarene swine bent on hurling themselves into an abyss: 'This is the road to chaos and barbarism. Virtually every necessity or convenience of life is capable of being objected to by somebody with an axe to grind. But how much civilisation is left once people decide to disobey any laws they disagree with, and reserve the right to prevent others going about their lawful affairs because they have taken exception to them?'

This rather misses the point. For those who have opposed the road building programme for many years, either passively at a distance or through direct involvement in protest organisations, it can come as a blessed relief to hear someone cutting straight through all the rationalisations offered by Government and road builders simply by declaring: 'This land is sacred, it matters to local people and to us. Future generations have the right for it to matter to them too. So take your bulldozers and your chainsaws, your cost-benefit analyses and impact assessments, and get lost!'

It is the raising of that uncompromising, values-driven voice that is almost certainly more important to the environment movement than the actions themselves. The unavoidable truth is that no major scheme has ever been stopped once the first sod has been turned by the contractors involved. But the Government's roads programme has been cut back and back *in part* at least because of the increasing influence of anti-roads campaigners, and some specific schemes (the East London River Crossing, for example, which would have sliced through Oxleas Wood) withdrawn as a direct result of the threat of civil mass disobedience. Beyond that, these policy shifts are as much the result of years of academic research and patient foot-slogging along the corridors of power as of high-profile direct actions, leading to an extraordinary level of interdependence between organisations for all that they are using highly divergent tactics.

The New Pluralism

The environmental NGOs, therefore, remain central to the future of the environment movement, but in a very different way to the seventies and eighties. Perhaps the biggest single change between the old environment movement and the new lies in the number and relative influence of participating sectors or interest groups.

When I went to Friends of the Earth in 1984, the positioning of this front line organisation was very simple. It dealt principally with the government and sought members and money in order to increase pressure on government. It was the same with most environmental organisations at that time. To a quite extraordinary degree, the environment agenda was brokered by the government and the NGOs between them, with other sectors responding to that agenda, the media commenting on it reactively, and academics picking up research contracts (if they were lucky) off the back of it.

The scene confronting the Director of Friends of the Earth today is quite different. There are now several principal players (including the business community and local government), as well as a huge supporting cast of academic disciplines, professional associations, community groups, educationalists, religious and spiritual initiatives, and so on.

The upshot is that the environment agenda no longer belongs to the environmental NGOs; the movement today is 'owned' by a multitude of different agents in society, each with their own view on how best to move things forward. Since shared ownership and the much wider dissemination of knowledge and concern about the environment were always key goals of earlier environmental campaigners, you might suppose that this development would be a cause for celebration. In fact, many NGOs have reacted with suspicion as to the true motives of these new players, and with considerable apprehension at the possibility that they might face co-option rather than co-operation.

This is particularly the case with the business community. Since the late eighties, there has been a deeper and deeper engagement in the environment agenda by individual companies and organisations representing business interests in general. To start with, much of it was indeed pretty flaky, motivated more by the desire for cheap PR than by any genuine intent to improve environmental performance.

A decade on, however, there is no denying the substance of that engagement; driven by a swathe of new legislative pressures, fearful of the loss of reputation that comes from being branded an environmental pariah, and increasingly aware of the very real opportunities of doing good business by doing it in a leaner, greener way, the environment is now a serious issue for a very large number of the very largest companies.

This has led several commentators to conclude that we are now entering a 'third phase' of the environment movement, after a first phase characterised by outright denial on the part of governments and businesses, necessitating full-frontal confrontation on the part of NGOs, and a second phase marked by growing regulatory pressure from governments and higher expectations on the part of both consumers and voters. This left companies with little choice but to comply, albeit on their own terms. Only now is it possible to envisage a genuinely committed and constructive contribution from the business community, with companies often moving well beyond the minimum required by law.

The prospect of partnership of that kind is exciting, and NGOs are starting to explore this new territory more systematically than ever before. The emergence of organisations like EPE (European Partners for the Environment) and Forum for the Future (which campaigns for the *answers* to the problems through working partnerships with leading businesses, local authorities and universities), together with a host of government-convened initiatives (such as the President's Council on Sustainable Development in the United States and our own Round Table in the UK) are useful indicators of the way this trend may develop.

Yet caution is still an important prerequisite for pursuing this 'solutions agenda'. A spate of new books (such as David Korten's *When Corporations Rule the World*, Andrew Rowell's *Green Backlash*, and Rifkin and Goldsmith's *The Case Against the Global Economy*) demonstrate all too clearly how foolish environmentalists would be to suppose global capital has been turned overnight into a benign force, or that the majority of transnational corporations have done anything more than wake up to the fact that there is a bit of a problem out there with the global environment.

It all makes for a very interesting scene! The 'dark Greens' are understandably worried at the prospect of this third phase, fearing the wholesale erosion of the values and philosophical insights that have underpinned the work of the environment movement since the publication of Rachel Carson's *Silent Spring* in 1962. The 'red Greens' understandably pour scorn on the political naivety of the environment movement, deploring the ideological vacuum at the heart of most NGOs, and asking (as Ted Benton does in this collection) how it can be that, at a time when capitalism is more nakedly exploitative and environmentally destructive than ever before, so many environmentalists should be throwing in their lot with the agents of that destruction.

The more pragmatic 'light Greens' ask, equally understandably, what choice we now have other than to work with different sectors, and to deepen their emerging commitment rather than reject it out of hand, to take the risk inherent in any partnership process rather than lose all in sterile isolationism.

It is not an easy nut to crack. *If* that stream of innovative creativity amongst entrepreneurs, financiers and engineers could be channelled into producing the goods and services we need on a genuinely sustainable basis; *if*, as Robin Murray suggests, we could move rapidly beyond the rather facile business of 'greening up' the existing economy into the business of alternative business *systems*, working through the market but not principally for the market; *if* governments could be persuaded to promote the development of the social economy (or the 'third sector', as it is sometimes called, operating between the public and private sectors), building capacity at the local level to act as a bulwark against the vicissitudes of a global economy that will find it harder and harder to generate the jobs even as it goes on generating the growth—if all that could emerge as a consequence of working with rather than against the energy of the business community, then the benefits could be enormous. If not . . .

Bringing out the Values

As an active protagonist in this debate, it sometimes takes on the appearance of a battle between warring Trojan Horses. 'Our side' likes to think that we have successfully infiltrated core concepts such as eco-efficiency and genuine sustainability into the very heart of the industrial citadel, mitigating its destructive power and threatening to subvert it altogether at some stage in the future. Critics argue, by contrast, that 'their side' is successfully infiltrating fundamental tenets of global capitalism (such as the need for pricing all environmental goods and services, and that good old oxymoron 'sustainable growth') into the leafy glades of the environment movement. As ever, potentially irreconcilable value systems are at work here, manifesting themselves in heated discussion about different political strategies.

By and large, most environmental organisations fight shy of making those values more transparent. In sharp contrast to the animal rights movement, the underlying premises behind any campaign or policy stance often remain invisible. We might, for instance, all agree that we need 'to save the rain forest', but why? To regulate global climate patterns? To protect endangered species/indigenous peoples/future generations? To prevent the loss of genetic resources of potentially incalculable economic benefit to us?

It has long been 'a given' of the modern environment movement that it is highly critical of the *instrumental* use of nature that has been at the heart of the industrial revolution, and that it would like governments and people to recognise and respect the *intrinsic* value of all other species, regardless of their potential usefulness to us. This philosophical bedrock has always permitted environmentalists, when it suits them, not to be bound by the soulless techniques of cost benefit analysis and 'contingent valuation', and to defend certain courses of action in defence of the natural world that might very well have 'no economic rationale'.

That bedrock still remains intact, but there is growing cause for concern that immersion in the discipline of sustainable development is insidiously weakening the anti-instrumentalist mind set. As far as most economists and politicians are concerned, sustainable development has come to mean in practice the management technique of sustainable utilisation: how can we maximise the use of a particular asset without destroying its capacity for self-regeneration? That is a perfectly reasonable question to ask, but not if it is the *only* question to which an answer is being sought. As far as mainstream politics is concerned, it is.

Robin Grove-White has suggested that this is an unavoidable consequence of NGOs being drawn in by the way government and conventional economists choose to 'frame' or define the environmental problem. The co-option that we should be worrying about is not so much that which is inherent in compromises on policy, or even on tactics, but rather the co-option of our entire value system.

Have we, for instance, been co-opted in the way in which environmental organisations have come to use the notion of 'sound science' to validate and justify their campaigning positions? Way back in the seventies, the science of it all really did not count for very much; conveying the broad picture (that a finite natural system cannot sustain the infinite expansion of one species at the expense of all others) did not seem to demand absolute scientific rigour, though I suspect we paid a higher price for that than we were prepared to acknowledge.

By contrast, the eighties saw an ever greater emphasis put on the importance of 'getting it right scientifically', if we were to make any impact at all on Government, the media or even the general public. NGO research budgets grew rapidly, as did working relationships with the universities and freelance academics. 'Scientific credibility' became one of the most important elements in NGOs' mission statements.

But what if that science has been shaped by 'prior social commitments' (as Brian Wynne argues), or worse yet (as Ted Benton argues), itself co-opted by global capitalism so that its principal purpose is now to extend control over the natural world and emerging nations?

Personally, I am not so sure that this is as big a problem for the environment movement as some would have us believe. As Bob Worcester's polling evidence on 'confidence in scientists' would seem to demonstrate, there is no profound turning away from the authority of science itself, but rather from certain interpreters of that science. As evidenced by the highly creative debate around concepts such as the precautionary principle, the environment movement is peculiarly well placed to go on arguing the importance of sound science, while simultaneously arguing that even the soundest of science should not necessarily be presumed to provide all the answers. To conclude, however, that further investment in science is 'not part of the solution' would seem to be a recipe for absolute despair.

Environmentalists could certainly afford to be raising questions about the limitations of science (and indeed of scientific materialism itself) far more vociferously than before. Yet the values debate remains surprisingly muted in most environmental organisations, with very little interaction between activists and the increasingly lively discipline of environmental ethics. 'Deep ecology', which has had such a huge impact on the environment movement in the United States, has next to no currency here in the UK, and overtly spiritual or religious approaches are still considered suspect or positively dangerous by the majority of environmental organisations.

Calls to keep the values/spiritual side of things separate from the political and economic side sound increasingly dated and unworldly. Try suggesting that to an anti-roads campaigner perched high up in a tree scheduled for felling, or to those who feel so passionately about stopping the export of live animals to the continent.

As I have already suggested, the success of the environment movement over the last twenty-five years can in part be attributed to its explicit rejection of prevailing philosophical orthodoxies. The motivation to get involved in environmental issues has always been driven partly by external political factors, and partly by internal 'existential' factors. If we are to ensure that the environment movement does not 'lose its soul', does it not become ensnared by the seductive power of industrial attitudes it seeks to transform, then its values need to be brought to the forefront of its work rather than allowed to languish in obscurity.

Sustainable Development and Social Justice

The other great challenge the environment movement must meet if it is to maximise the potential of the partnerships now on offer, without losing its soul, is *explicitly* to address the whole question of social justice. Or, to put it another way, to remember that sustainable development and environmentalism are *not* the same thing, and that the 1992 Earth Summit rang down the curtain on that kind of orthodox, old-world environmentalism which took no account of distributional or social justice issues.

Unfortunately, the all-important linkages between environment and social justice first mapped out in *Our Common Future*, the 1987 report of the World Commission on Environment and Development which argues 'sustainable development requires meeting the basic needs of all, and extending to all the opportunity to fulfil their aspirations for a better life; a world in which poverty is endemic will always be prone to ecological and other catastrophes' have actually become less rather than more salient since 1992. In the intervening five years, the total volume of aid declined from around $65 billion per annum to under $55 billion today, and additional resources specifically promised to assist developing countries to meet the challenge of sustainable development have been few and very far between.

The consequences of that process have been starkly mapped out in the very powerful Human Development Report published by the UN Development Programme in 1996. Richard Jolly, the chief author of the report, pointed out that, whereas the richest 20 per cent of the world's population was thirty times better off than the poorest 20 per cent in 1966, they are now sixty-one times wealthier. Perhaps the one statistic that really does concentrate the mind in all of these discussions is the fact that the wealth of the world's 358 billionaires is equal to the combined annual incomes of around 2.2 billion people.

In a splendidly provocative article that sets the tone for the whole report, Nobel economics prize-winner Robert Solow goes to the heart of the environmental and social justice issue: 'Those who are so urgent about not inflicting poverty on the future have to explain why they do not attach even higher priority to reducing poverty today.'

Yet it can easily be demonstrated that social justice issues (both in the UK and internationally) have played little part in the output of environmental NGOs since the Earth Summit. Again, only Friends of the Earth would seem to have seized hold of this challenge strategically. It has quietly built up an impressive sustainable development department, winding down some of its more specific and conventional environmental campaigns, and its two most significant reports in 1996, for instance, dealt with the issues of health and employment.

This is an agenda that has been strangely absent in the UK, despite compelling evidence that many environmental disbenefits fall disproportionately on poorer people. In October 1995, for instance, the National Asthma Campaign published a survey showing that asthma is a far more disabling disease for poorer social groups. Eighty-six per cent of those disabled by asthma are in social groups 3, 4 and 5, with only 14 per cent in social groups 1 and 2.

In stark contrast to the United States, there is no real 'environmental justice' movement in this country. That may now change with the arrival of Real World, a coalition of more than forty-five campaigning NGOs launched in April 1996 specifically to help force sustainability issues on to the agenda of the general election. Its membership embraces not just the predictable environmental and international development agencies, but organisations like the Poverty Alliance, Church Action of Poverty, and the Public Health Alliance.

It is not always an easy fit. Many environmental organisations are uneasy that it has proved almost impossible, against a backdrop of widening disparities in income, to confront the whole issue of overconsumption in countries like the UK. Just so long as the discourse remains fixed in the reassuring domain of more ethically-responsible, environment-friendly consumption, it is all sweetness and light. Head off into the territory of *less* consumption and down come the shutters.

Is this a further compromise on the part of environmental campaigners, with questionable 'politically correct' overtones? Almost certainly, but justified in this instance by the overwhelming need to cajole the broad environment movement into a more politically engaged orientation, especially at a time when this nation's redistributive instincts seem to be ticking over at a very low level indeed.

This situation may change much sooner than New Labour's continuing swing to the right would seem to indicate. Many in the environment movement remain confident that the onset of the Millennium will promote amongst the general public certain tendencies which have always been central to an environmental ethic: the defence of public goods, anti-individualism, the recognition of value beyond money, the intrinsic value of the natural world, international solidarity and so on.

This may, of course, be just the latest manifestation of the environment movement's perennial ability to allow hope to triumph over adversity. But even the most cursory comparison of the state of the movement today and its situation ten, let alone twenty-five years ago, makes some of the prevailing mood of doom and gloom look just a little self-indulgent. For all the dilemmas it is now wrestling with, many of the environment movement's old strengths (diversity, tactical flexibility, credibility, and an intuitive resonance both with people's hopes and their fears for the future) are now reinforced by its new-found inclusivity and pluralism, and a growing political maturity.

Biographical Note

Jonathon Porritt is Director of Forum for the Future, a new charity which campaigns for solutions to today's environmental problems. He is a former Director of Friends of the Earth.

Ecological Modernisation: Restructuring Industrial Economies

ANDREW GOULDSON and JOSEPH MURPHY

Introduction

Historically, the relationship between economic development and environmental protection has been seen as one of mutual antagonism. Those who have been primarily interested in the performance of the economy have generally perceived environmental protection to be a brake on growth. Conversely, those who have been principally concerned about the quality of the environment have tended to see economic development as the root of the environmental problem. It can be argued that the consequent conflict between industrialists and environmentalists has defined the climate of environmental politics to such a degree that the level and nature of environmental policy making have been severely constrained.

This conflict is clearly justified in some instances. But in recent years the concept of 'ecological modernisation' has been developed to try to move beyond it.[1] Ecological modernisation proposes that policies for economic development and environmental protection can be combined to synergistic effect. Rather than seeing environmental protection as a brake on growth, ecological modernisation promotes the application of stringent environmental policy as a positive influence on economic efficiency and technological innovation. Similarly, rather than perceiving economic development to be the source of environmental decline, ecological modernisation seeks to harness the forces of entrepreneurship for environmental gain. Thus, ecological modernisation suggests that economic and environmental goals can be integrated within the framework of an advanced industrial economy.

The Nature of Ecological Modernisation

The basis for this is the recognition that economic growth need not automatically lead to higher rates of environmental degradation. By increasing the *environmental efficiency* of the economy—the rate of environmental damage caused per unit of output—growth can be achieved even while environmental damage is reduced. This can occur in three different ways. Polluting materials can be substituted by others more environmentally benign: fossil fuels by renewables, for example. The use of materials can be made more efficient, for example through waste minimisation and recycling. And the composition of

© The Political Quarterly Publishing Co. Ltd. 1997
Published by Blackwell Publishers, 108 Cowley Road, Oxford OX4 1JF, UK and 350 Main Street, Malden, MA 02148, USA

output can change, away from material intensive products to those with lower environmental impact.

The theory of ecological modernisation argues that such improvements in environmental efficiency need to occur at two different levels. First, ecological modernisation seeks structural change at the *macro-economic* level. It looks for industrial sectors which combine higher levels of economic development with lower levels of environmental impact. In particular, it seeks to shift the emphasis of the macro-economy away from energy and resource intensive industries towards service and knowledge intensive industries. It argues that changes in infrastructure and technology—such as through public transport provision, land use planning and the use of information technologies—can also make a major difference to environmental impact. Thus through a combination of sectoral, infrastructural and technological change, it proposes that the structure of the macro-economy should be reoriented to establish a more environmentally benign development path. This path would require the consumption of fewer resources and the generation of less waste, while also creating employment and improving economic welfare.

Secondly, ecological modernisation assigns a central role to the invention, innovation and diffusion of new technologies and techniques at the *micro-economic* level. In particular, it seeks a shift away from traditional 'control' technologies towards the development and application of 'clean' technologies and techniques. Control technologies (for example, effluent treatment plants) are end-of-pipe additions to products or processes which capture and treat waste emissions in order to limit their ultimate impact on the environment. By contrast, clean technologies, such as energy efficient light bulbs and low-pollution manufacturing systems, integrate environmental considerations directly into the design and application of products and processes. In this way they avoid or reduce the environmental impact at source. Clean techniques, such as environmental management systems, can have the same effect.

Advocates of ecological modernisation argue that such changes, at both macro-economic and micro-economic levels, have the potential to make significant improvements in the environmental performance of industrial economies. Ecological modernisation is thus presented as a means by which capitalism can accommodate the environmental challenge. Rather than environmental protection being a threat to capitalism, it is seen as a spur to a new phase of capitalist development. This has led to criticism amongst some environmentalists, who foresee the co-option and neutralisation of the environmental movement.[2] Ecological modernisation is silent on crucial questions of social change—for example concerning social justice, the distribution of wealth and power and society-nature relations. It is thus simply an attempt, critics warn, to legitimise and sustain the very structures and systems that have been responsible for environmental decline.

On the other hand, as its advocates point out, this can also be seen as an argument in its favour. By accepting capitalism, but seeking to reform it in technologically and politically feasible ways, ecological modernisation at last provides some purchase for environmentalists in mainstream economic debate.

Integrating the Environment and the Macro-Economy

Ecological modernisation is generally presented as a conceptual framework for understanding the potential of economy-environment integration. But to some extent, at least, ecological modernisation appears actually to be occurring.

The evidence for this comes from an empirical study on the relationship between economic development and environmental impact in 31 countries during the post-war period (up to the mid-1980s).[3] This reveals that there has been a 'decoupling' of economic growth from environmental damage. Many countries have seen improvements in macro-economic environmental performance. The environmental intensity of production in certain key sectors has gradually fallen, measured both in terms of resources consumed and emissions generated. Moreover this seems to be closely related to changes in the technological and sectoral composition of these countries' economies over the same period.

But this general trend towards technological and sectoral change in industrial economies is not uniform across countries. Both the nature and the extent of the decoupling of economic development from environmental impact varies. Some countries, such as Denmark, France, Germany and the United Kingdom, have experienced an absolute decoupling. In these cases, in certain key sectors, economic output has increased and environmental impact has simultaneously decreased in absolute terms. Others, such as Austria, Finland, Norway and Japan have experienced only a relative decoupling of economic output and environmental impact. Although each *unit* of output in the key sectors is associated with a lower environmental impact than before, these gains in environmental efficiency have been more than offset by additional impacts arising from an expansion of output. Finally, a minority of countries, mostly the former command economies of Central and Eastern Europe, have experienced a negative relationship between economic output and environmental impact: as growth occurred, environmental damage rose too.

In those countries that have realised an absolute or relative improvement in the relationship between economic growth and environmental impact, this is associated with a general shift away from structures emphasising the *volume* of production towards those emphasising its *value*. This involves a decline in the absolute or relative importance of energy and resource intensive industries in favour of an increase in the importance of the service and knowledge intensive industries. This structural change has apparently occurred 'unintentionally', without specific and deliberate policy intervention. This had led the study's authors, Martin Jänicke and his colleagues, to refer to 'an environmental *gratis* effect'. It appears that advanced economies have begun to experience the effects that would be associated with a programme of ecological modernisation without even trying.

Three caveats however need to be attached to this conclusion. First, much of the sectoral shift has occurred not because of a decline in the *consumption* of manufactured goods in advanced economies, but because their *production* has been relocated over time to newly industrialising countries, particularly in East Asia. Development has been spatially uneven and patterns of trade have changed. In such instances some countries will give the impression statistically of having improved their environmental performance when in fact, intentionally or unintentionally, they have merely externalised their impacts by importing rather than producing certain goods. It is likely, therefore, that the improvements in the environmental performance of advanced countries have only been realised at the expense of deteriorations in developing countries. Environmental degradation has effectively been 'exported'. If this is true, it is not clear whether the 'gratis' effect is only possible in a select number of post-industrial economies; it may not be possible for all countries simultaneously to secure an improved relationship between macro-economic development and the environment. Secondly, it is important to ask whether the benefits offered by the 'gratis' effect have been sufficient. In many countries, as noted, the benefits arising from an improvement in macro-economic environmental efficiency (falling environmental impact per unit of output) have been overcome by continued growth in output. Overall damage has therefore risen. Even more importantly, even in those countries where the reduction in environmental impact has been absolute, it has been small, and has certainly fallen short of what is required for long-run sustainability. Damage continues to be done to the environment, even if consumption and emission rates are somewhat lower than they used to be. Thus energy consumption may have fallen, but industrial economies are still producing more carbon dioxide than the atmosphere can absorb without global warming. Urban air quality has improved since the 1940s and 1950s, but it is still damaging to health (and it is now getting worse). Any ecological modernisation that has occurred so far has not been on a sufficient scale to back up the claims of its advocates that it offers a genuine escape from the environmental problem.

Now of course it may be possible for the ecological modernisation process to be speeded up or expanded by active government policy. But the third caveat is that on the basis of past experience we don't yet know whether government intervention can actually serve to promote overall environmental improvement. To date the environmental benefits of macro-economic structural change have been secured largely by accident. It may be that ecological modernisation is therefore exogenous rather than endogenous in influencing policy. That is, it may simply occur through the 'natural' evolution of economic development, and be unresponsive to deliberate intervention. Evidence is beginning to emerge, however, which suggests that this is not the case. Recent economic modelling exercises indicate that policies are available which could make significant improvements to the relationship between economic development and environmental protection. An important example is a study for the European Commission examining the potential benefits of integrating environmental and economic policies.[4] The study projects European Union (EU) economic performance under various environmental policy scenarios.

The first scenario represents a continuation of recent economic trends and existing environmental policies. Under this scenario there is a significant deterioration in environmental quality in the period to 2010. Carbon dioxide emissions, for example, are projected to rise by 30 per cent and solid waste by 30 per cent. This is the 'business as usual' position. The second scenario allows for the implementation of new environmental policies currently under consideration at the EU level, including a carbon/energy tax. Under this scenario notable environmental benefits are realised in some areas. But these benefits are secured only with a significant negative impact on economic growth.

The third scenario assesses the implication of fully integrating environmental objectives into macro-economic and sectoral policies. It projects the widespread application of taxes and other fiscal instruments on energy, transport, industrial pollution and water abstraction, along with the establishment of support mechanisms for the research, development and rapid diffusion of clean technologies. The revenues from these taxes are used to reduce social security contributions paid by employers, thereby lowering the cost of labour. Under this scenario, representing a major environmental economic programme for Europe, highly significant environmental benefits are achieved: reductions in air and water pollution and solid waste of between 50 and 70 per cent, a positive enhancement of biodiversity, and a stabilisation (though not, still, a reduction) in carbon dioxide emissions.

What is particularly significant is that these benefits are not achieved at the expense of economic performance. Growth is in fact slightly higher than in the 'business as usual' (base) scenario. Even more important, employment rises. The integrated package of environmental policies is projected to raise European employment levels by 15 per cent or over 2 million people by 2010. This occurs through shifts in the sectoral composition of the economy: as the relative prices of employment and energy/pollution change, the economy gradually becomes more labour and service intensive and less resource intensive. That is, this scenario projects a significant process of ecological modernisation.

Integrating the Environment and the Micro-Economy

Experience at the micro-economic level mirrors that at the macro-economic level. The widespread application of control technologies for pollution abatement has reduced environmental impact per unit of output in many sectors. But this has often been outweighed by the growth in output, and environmental damage, particularly in key industrial sectors such as chemicals, remains very substantial. To date the take-up of 'clean' technologies and techniques remains very limited. Investment in clean production systems has been confined to a relatively small number of large companies operating in industrial sectors which are exposed to particular pressure to improve their environmental performance. It has not been universal even within these companies. Amongst smaller companies and those in less heavily regulated sectors and countries there has been very little investment in this field.

Yet as a considerable body of evidence now shows, there is clearly the potential for clean technologies and techniques to make major improvements in firms' environmental impact. Even more importantly, a number of studies have established that this improved environmental performance is associated with significant benefits to the firm. A Dutch project, for example, sought to establish the costs and benefits associated with various options for improved environmental performance.[5] These options included both minor changes in the technologies and organisational practices of companies and major innovations in their products and processes. For the ten participating companies, the study identified one hundred and sixty-four environmental management options, of which forty-five were immediately implemented, sixty-six were implemented in the medium term and the remainder were explored through on-going research and development. Of the forty-five options implemented, twenty generated cost savings and nineteen were cost neutral. A range of indirect benefits, relating for instance to improvements in product quality, were also realised.

The common finding of this and other similar studies has been that clean production systems offer both environmental and economic benefits to the firm. Yet their potential remains largely unexploited.

Barriers to Change

It would seem that at both the macro-economic and the micro-economic levels the relationship between economic development and environmental protection could be significantly improved. Yet progress to date has either been accidental or has been restricted to a small number of countries or companies. Why? What are the barriers to ecological modernisation?

At the macro-economic level, this question is concerned primarily with government policy. The advanced countries have been implementing concerted environmental policy for a quarter of a century. But it has evolved in a process based not on strategic thinking but on short-term reaction to crisis or failure. Over time, the emphasis of policy has gradually shifted from the effects to the causes of pollution. But while changing approaches have undoubtedly had an impact on the rate of environmental decline, policy continues to be largely reactive rather than proactive in nature. As modern industrial society becomes increasingly complex, it is widely argued that this is no longer adequate. The social and ecological risks associated with technological progress are too pervasive. Thus it is increasingly argued that more proactive forms of environmental policy are needed if further reductions in the level of environmental quality are to be avoided. As we have seen, such policy can offer economic as well as environmental benefits.

The theory of 'state failure' offers an explanation of why—despite its potential economic benefits—this has not yet occurred.[6] Alliances between government and industry constrain the search for new approaches to policy. Both government and industry tend to favour reactive and standardised approaches to policy, even though proactive and flexible solutions are available which are both more efficient and more effective. For government, standardised solutions ensure that their own costs of policy design and delivery are minimised. For industry, standardised approaches provide a clear and readily understandable framework which in theory imposes an equal burden on all regulated companies (the 'level playing field'). Reactive approaches are relatively easy to accommodate as they do not challenge existing strategies and systems. Over time, for both government and industry, organisational and technological allegiances to existing policy approaches then help to ensure that these continue. Thus, policy becomes 'locked-in' to a reactive and standardised approach even though more proactive policies are available and might offer economic and environmental advantages.

Barriers to change also exist at the micro-economic level. Even where there is potential for economic benefit from environmental improvement, the scope for realising this potential is commonly restricted by the nature of the innovation process. Limited information, shortages of managerial capacity, the unavailability of financial capital and risk aversion can all inhibit the rate of innovation in new technologies and techniques.

These barriers to the adoption of new technologies and techniques are compounded by factors which constrain the direction of any innovative activity that does take place. New innovations depend upon a system or network of organisational, strategic and other relations within a firm or industry without which their adoption would be impossible. Yet they must be introduced into systems that have generally been developed for or adapted to older technologies and techniques. Their introduction may therefore require changes to parts of a pre-existing system to ensure compatibility with the system as a whole. This process of change may encounter considerable resistance. Firms that have successfully mastered the operation of old technologies and techniques commonly find it difficult to overcome the limits of existing expertise to acquire and assimilate the skills and knowledge needed to apply new technologies and techniques successfully. Thus an innovation which requires only incremental change to existing systems is more likely to be adopted than one which requires more radical change. Consequently, the development of new technologies and techniques tends to take place within particular trajectories in an incremental and evolutionary form. This inhibits the development of more radical innovations, of which the most dramatically environment-improving clean production systems tend to be examples.

The ability of a new technology or technique to influence an existing system is affected over time by the dynamic and self-reinforcing impacts of the so-called 'scale' and 'learning' effects.[7] As an innovation is diffused throughout an industry its costs decline as production expands and economies of scale are realised. This scale effect is complemented by a learning effect: as experience with the production and use of the innovative accumulates, its quality increases. Information and knowledge gathered by actual users is transferred to potential users so that the uncertainty associated with the application of a new innovation is decreased. The increased community of users facilitates joint problem solving and cumulative learning. It also encourages factors such as the availability of an appropriately skilled labour force and the provision of training and maintenance services. All of these factors increase the efficacy and efficiency of the innovation as it is diffused, so that rather than encountering diminishing returns, innovations commonly experience a period of increasing returns to adoption.

These scale and learning effects have two contradictory results for the take-up of new technologies and techniques. Once an innovation is established, they help it to develop further and to be diffused more widely. But whatever their apparent potential, innovations which have yet to benefit from the scale and learning effects are inhibited. They may be perceived to be expensive, incompatible, unproven and without the necessary support mechanisms. This creates something of a paradox. Potentially beneficial innovations such as clean production systems may not be selected due to their initial inefficiency, while the only way that they can become efficient is by being selected. Both the encouragement and inhibition of innovation are manifestations of the 'lock-in' which firms experience to existing technologies and production systems. It would seem that overcoming the lock-in effect, in order to establish clean production systems more widely throughout an economy, will require active intervention in the innovation process.

The Need for New Policy Approaches

It is apparent that existing approaches to environmental policy fail to establish either a sufficient incentive or an adequate imperative to overcome the barriers that restrict change at both the macro-economic and the micro-economic levels. New approaches to policy are therefore required if existing economic development trajectories are to become more environmentally benign—if ecological modernisation, that is, is to be encouraged.

At the macro-economic level, effective environmental protection can only be achieved through a realignment of broader policy goals. A continued reliance on environmental policies which can only react to the negative impacts of other policy areas is both ineffective and inefficient. Instead environmental objectives need to be integrated into previously separate policy areas such as industry, energy, transport, land use planning and trade. This integration should ensure that the impacts of policies concerned with economic development reinforce rather than undermine the efforts of policies associated with environmental protection. Such policy integration, which is a central component of proposed programmes for ecological modernisation, is likely to demand a significant degree of institutional restructuring and capacity building. The boundaries, objectives, functions and cultures of government institutions concerned with both environment and economy will need to be redefined.

Institutional change will need to be accompanied by an exploration of alternative and innovative policy measures. The integration of environmental objectives into fiscal policy is one area of particular potential. In recent years it has been widely argued that existing fiscal structures tend to encourage the over-exploitation of environmental resources and the under-exploitation of labour resources. Consequently, a realignment of fiscal incentives and disincentives—the so-called 'ecological tax reform' discussed by Stephen Tindale in this volume—could encourage a shift towards a more environmentally benign, more labour intensive macro-economic development path.

While the reform of fiscal structures at the macro-economic level would certainly contribute to improved environmental protection in some areas, at the micro-economic level the technological and organisational factors discussed above are likely to constrain the technical change that it would stimulate. These factors are likely to limit the extent to which actors are able to respond to the revised set of incentives and disincentives that a programme of fiscal reform would establish. New policy approaches at the micro-economic level will therefore also be necessary.

Recent research has shown that regulation can itself be a stimulus to innovation.[8] Regulatory pressure—often in anticipation of new regulations and taxes, as well as in the period following their introduction—pushes companies into exploring new production methods which will reduce costs. This is an important insight in itself, countering the traditional (and still widespread) view that regulation is necessarily bad for competitiveness and investment. But it is also now becoming clear that the *styles* and *structures* of regulation can affect both the nature and the rate at which innovation occurs, in particular by overcoming the barriers and constraints discussed above.

Research has shown that a 'hands-on' regulatory style has particular benefits.[9] By this is meant the use of expert regulators interacting with regulated companies in a flexible, intense and cooperative way. The hands-on approach to the implementation of regulation helps companies to overcome the barriers to innovation by transferring formal knowledge and tacit understanding about the various options for environmental improvement. Importantly, it also encourages companies to extend their search for a response to regulation beyond those control technologies that are expedient in the short term. It helps them, that is, toward those clean technologies and techniques that offer improved economic and environmental performance in the medium to long term. Hands-on approaches to implementation, it has been found, can actually catalyse a 'culture change' in regulated companies so that they become more aware of the potential benefits associated with clean technologies and techniques, more committed to their development and application, and more able to innovate in the future.

Hands-on approaches to the implementation of regulation are not alternatives to economic instruments and legal standards. They complement the incentive and statutory effects of such instruments by engaging directly with firms as they respond to these. They therefore help to establish the conditions under which existing policy instruments become more effective. In economic terms, they raise the elasticity of response, encouraging a larger improvement in environmental performance from the same set of instruments. Changing the regulatory style in this way is therefore likely to be an important component in any strategy for ecological modernisation.

Conclusion: Ecological Modernisation in the Global Economy

The concept of ecological modernisation proposes that advanced developed economies can be restructured in favour of the environment. They can be shifted onto a different development path, in which absolute levels of environmental damage are significantly reduced, even while growth and welfare are maintained. There is evidence at both the macro-economic and micro-economic scales that such a shift is technically possible. But it will not happen automatically. It will require a new approach to policy on the part of governments. At a macro-economic level this will require the integration of environmental and economic policy making, involving significant institutional learning and reform. A reorientation of the fiscal base towards the environment and away from labour is likely to be an important part of this. At a micro-economic level the crucial need is to overcome the barriers to innovation in firms. In addition to more stringent regulatory standards and economic instruments, this will require a hands-on regulatory style.

Ecological modernisation may be theoretically feasible, but how likely is it to occur in reality? As we have seen, it is not actually happening yet on any significant scale. But other profound economic changes are.

Under the impact both of new information technologies and liberalised trade and financial regimes, the processes of economic 'globalisation' are now affecting all advanced economies. As capital has become more mobile, investment is increasingly international. The dramatic economic growth of the newly industrialised countries, particularly in East Asia, has transformed the global location of manufacturing, and increased the international competitive pressures felt by advanced economies. This has led in turn to a marked shift in the public policy agenda in Europe, away from the 'social model' of labour market protection and welfare spending. It has been increasingly argued—particularly, but not only, in the UK—that global competition demands a lower-cost model of labour market flexibility, reduced public spending and a general drive to deregulation in industrial policy.

At the same time, the processes of production have been changing. New technology and new forms of firm organisation have created the so-called 'post-fordist' era.[10] Mass production of standardised goods and services has given way to 'flexible specialisation', batch production of differentiated products for niche markets. New technologies and organisational techniques have been developed, such as 'just-in-time' stock control and delivery and instant transmission of sales information from shop to factory. New networks of firms have been created, with larger corporations contracting out production to smaller firms, and vertical lines of control being established in many industrial sectors, from raw material production to retail.

All these trends have been powerful, generating a major restructuring of industrial economies over the past twenty years. In this context, what prospect is there for a different kind of restructuring, an ecological one, to be promoted at the same time?

The portents look mixed. On the one hand, the climate of increased global competition is not conducive to the imposition of new regulatory regimes. The present momentum is all in the other direction, towards deregulation and lower costs. In so far as transnational companies' efforts are focused on investment in the newly industrialising countries and on selling to their rapidly expanding markets, environmental pressures, which come almost entirely from the developed world, are relatively weak.

On the other hand, competitive pressure is an important spur to innovation. Innovating firms give themselves a 'first mover advantage' over their competitors. Flexibility and dynamism, the hallmarks of innovating firms, are widely regarded as important in the new competitive climate. Moreover, there would appear to be some compatibility between the industrial technologies and techniques of flexible specialisation and those of environmentally benign production. Both are based on the principle that costs should be cut by utilising resources efficiently, rethinking the flow of materials through the production process. 'Lean', that is, may also be 'clean'.

In truth, it has to be said that we do not know how far the current trajectories of globalisation and post-fordism might be compatible with ecological modernisation. But the appropriate response to this, surely, is to find out. The crucial lesson here is that a significant reorientation of the economy towards environmental efficiency, at both macro and micro levels, will require active government policy. In this sense proponents of ecological modernisation will find themselves a part of, rather than separated from, the crucial arguments about the future of the industrialised economies which are now raging in Europe and beyond.

On one side of the debate, as we have seen, there are those who argue that European economies will only succeed in the new global economy if they reduce the costs firms face, particularly through deregulation. On the other hand, advocates of a new post-Keynesian settlement argue that this is the route to disaster, as social standards are driven down to the lowest global denominator.[11] The advanced countries cannot compete with the newly industrialised economies on cost. Their success will therefore be built on higher value-added, the result of high levels of investment in new technologies and in human capital. This strategy must be supported by a greater, not a lesser, degree of regulation, both within Europe and in international trade agreements, in order to prevent low standards undercutting higher ones. Higher investment, particularly in economies such as the UK with historically low investment rates, requires active government policy to stimulate particular forms of demand.

These arguments, usually applied to labour and social standards, apply equally to environmental ones. High environmental standards create new markets in environmental technologies. These in turn offer significant potential for new specialist industries and sectors, with the possibility of substantial employment growth. But promoting such a development path will require active government intervention, both in the establishment of the higher standards and in support for firms to meet them. The new global economy does not have to be a deregulated one. The prospects for ecological modernisation would appear to depend on whether or not it will be.

Biographical Note

Andrew Gouldson is a lecturer in the Department of Geography at the London School of Economics and editor of *European Environment*. He is co-author (with Joseph Murphy) of *Environmental Policy as Practice: The Implementation and Impact of Industrial Environmental Regulation* (Earthscan, 1997).

Joseph Murphy teaches in the School of Geography and Earth Resources at the University of Hull.

Notes

1 See for example Gert Spaargaren and Arthur Mol, *Sociology, Environment and Modernity: Ecological Modernisation as a Theory of Social Change*, Wageningen, LUW, 1991. A similar concept of 'eco-restructuring' has been proposed by Robert Ayres *et al.*, eds., *Eco-Restructuring*, Tokyo, United Nations University Press, 1995.

2 See for example Maarten Hajer, 'Ecological Modernisation and Social Change' in Scott Lash, Bronislaw Szerszynski and Brian Wynne, eds., *Risk, Environment and Modernity: Towards a New Ecology*, London, Sage, 1994.

3 Martin Jänicke *et al.*, 'Economic Structure and Environmental Impacts: East West Comparisons', *The Environmentalist*, Vol. 9, Part 3, 1989.

4 DRI *et al.*, *Potential Benefits of Integration of Environmental and Economic Policies*, London, Graham and Trottman, 1994.

5 H. Dieleman and S. de Hoo, 'Toward a Tailor-made Process of Pollution Prevention and Cleaner Production: Results and Implications of the PRISMA Project' in

K. Fischer and J. Schot, eds., *Environmental Strategies for Industry: International Perspectives on Research Needs and Policy Implications*, Washington, Island Press, 1993.

6 Martin Jänicke, *State Failure: The Impotence of Politics in Industrial Society*, Cambridge, Polity Press, 1990; Andrew Gouldson and Joseph Murphy, *Environmental Policy as Practice: The Implementation and Impact of Industrial Environmental Regulation*, London, Earthscan, 1997. See also M. Jänicke and H. Weidner, eds., *Successful Environmental Policy: A Critical Evaluation of 24 Case Studies*, Berlin, Edition Sigma, 1995.

7 R. Kemp, 'An Economic Analysis of Cleaner Technology: Theory and Evidence' in Fischer and Schot, *Environmental Strategies for Industry, op. cit.*; L. Soete and A. Arundel, 'European Innovation Policy for Environmentally Sustainable Development: Application of a Systems Model of Technical Change', *Journal of European Public Policy*, Vol. 2, No. 2, 1995, pp. 285–385; OECD, *Technology and the Economy: The Key Relationship*, Paris, OECD, 1992.

8 See for example M. Porter and C. Van der Linde, 'Green and Competitive', *Harvard Business Review*, September/October 1995, pp. 120–33.

9 Gouldson and Murphy, *Environmental Policy as Practice, op. cit.*

10 Ash Amin, ed., *Post-fordism: A Reader*, Oxford, Blackwell, 1994.

11 See for example Will Hutton, *The State We're In*, London, Cape, 1995; Michael Jacobs/The Real World Coalition, *The Politics of the Real World*, London, Earthscan, 1996.

© The Political Quarterly Publishing Co. Ltd. 1997

Interpreting Sustainable Development: The Case of Land Use Planning

SUSAN OWENS

Introduction

As we approach the Millennium, sustainable development has become widely accepted as a policy goal. The British land use planning system, which has a long history of managing the environmental effects of social and economic change, has responded with enthusiasm to the new challenge. Planning is widely endorsed as a means of making progress towards sustainability, not least by the Government which sees it as a 'key instrument' in this respect.[1] Indeed, it is arguable that the *concept* of sustainable development has been adopted in planning more extensively, and more firmly on a statutory basis, than in any other field. Its interpretation in practice, however, remains problematic. The capability of the system to deliver sustainable development will certainly be tested at a time when growing demands for a wide range of goods and services–energy, housing, minerals, travel and water to name but a few—seem set on a collision course with rising environmental expectations. Some projections suggest, for example, that there could be a virtual doubling of aggregates output by 2011, an additional 4.4 million households by 2016 and a doubling of the amount of traffic by 2025—even more on rural roads.

One of the enduring messages of the Brundtland Report,[2] which popularised the concept of sustainable development in the late 1980s, was that economic growth and environmental protection could be compatible, indeed mutually interdependent. It is an acknowledged function of land use planning to promote such compatibility. In the real world, however, not all developments can be reconciled with all dimensions of environmental concern. What makes the planning system such an interesting example of policy practice is that it is not simply a technical means by which sustainability is implemented but an important forum through which it is contested and defined. In the process, enduring conflicts are not being reconciled by sustainable development as some had hoped and predicted; if anything, more fundamental contradictions are exposed. This chapter explores some of the tensions emerging as we struggle to determine what is sustainable 'on the ground'.

Published by Blackwell Publishers, 108 Cowley Road, Oxford OX4 1JF, UK and 350 Main Street, Malden, MA 02148, USA

Environmental Capacities and 'Balance'

One of the most fundamental tensions was noted by the House of Lords Select Committee on Sustainable Development, which found among its witnesses a 'critical distinction . . . between those who insist upon the need to establish and maintain a minimum environmental stock or capacity and those who accept the inevitability of trade offs between social and economic preferences and environmental resources'.[3] Many environmentalists and academics,[4] as well as statutory agencies like English Nature, believe that vital, irreplaceable or otherwise precious features of the environment (the ozone layer, for example, or ancient woodland) cannot simply be replaced by human-made capital (such as houses and roads) if we are to maintain or enhance overall levels of welfare. 'Critical environmental capital', they argue, merits special protection. Some go further, to suggest that while it is not possible to maintain all environmental assets in their current form, we should aim for no significant diminution of environmental quality over time. This is the 'constant environmental assets' rule, to be upheld by positive compensatory measures (such as landscape enhancement, or the re-creation of ecosystems) when non-critical assets are unavoidably lost or damaged. The need to protect what is critical and to maintain or enhance environmental quality defines the capacity of any given environment to accommodate change. Witnesses in the Select Committee's first category consider development to be sustainable if it takes place within environmental capacity constraints. Those in the Committee's second category offer a different interpretation, with the emphasis on integration and balance. They argue that there are few environmental assets so vital to human welfare that they should be removed from the arena of trade off, or they are sceptical about our ability to define environmental capacities, or both.[5] They tend to be optimistic about technological capabilities. From this second perspective, nothing is inviolate and sustainable development is to be achieved through a 'balance' of social, economic and environmental considerations.

These divergent, and ultimately incompatible, interpretations of sustainable development have important implications for land use planning in its attempt to reconcile the needs of development and environmental protection. In a capacity-constrained system, environmental limits would be defined in terms of critical environmental capital and the 'constant environmental assets' rule for any given geographical area. Development could then be accommodated to the extent that it did not cause unacceptable harm, depending in turn on the potential—conceptual and technical—to eliminate, mitigate or compensate for environmental damage. Since this potential is not generally fixed, limits to capacity do not translate in a deterministic way into limits to growth.[6] But on any typical planning timescale, and for critical assets, sustainability defined in terns of environmental capacity implies that there must be an environmental bottom line. It follows—since capacities established for particular localities could be aggregated—that there might be overall ceilings on certain types of economic activity at national level.

Not surprisingly, those whose main interests lie in development reject the concept of *a priori* capacity constraints and their wider implications. While accepting that the advantages of development might be outweighed by environmental considerations at specific sites, they see it as a key role of the planning system to achieve 'balance' between economic, social and environmental aspirations by providing for development in appropriate locations. Environmental quality should be protected 'as far as possible' but in the end it is negotiable when it comes into conflict with other goals: this has very different implications from the 'constant environmental assets' rule. The tension between these divergent interpretations of sustainable development is emerging clearly in both the policy and practice of land use planning.

Policy guidance and other government statements which might be expected to give a lead are ambivalent about the potential for trade off. The need 'to balance' competing demands for a finite quantity of land is seen as a key issue in the UK Strategy for Sustainable Development; but the planning system, according to the same document, can ensure that necessary development '. . . takes place in a way that *respects environmental capacity constraints*'.[7] Elsewhere the Major Government accepted that trade-offs between environmental and economic objectives may be inappropriate, not only in the case of 'major life or planet-threatening concerns', but sometimes 'for more modest or site-specific issues'.[8] Balance recurs as a desirable objective throughout the series of Planning Policy Guidance notes (which effectively set out Government policy as it is to be interpreted by local planning authorities) but a less utilitarian theme can also be traced. Certain environmental assets come close to being identified as critical and, in theory at least, not everything can be traded off against everything else: development which might damage a Special Area of Conservation (defined under the EU Habitats Directive), for example, can be justified only on grounds of 'overriding reasons of human health and public safety' or 'imperative reasons of overriding public interest'.[9] These and similar provisions for other protected areas are, in a sense, environmental capacity constraints, though they are not absolute and can often be undermined in practice by defining the public interest in narrowly economistic terms. What is missing from any policy statement, however, is the concept that environmental constraints might 'add up' to impose overall limits on particular forms of development and activity at regional or national level.

One example, concerning the vexed issue of minerals extraction, must suffice to illustrate how the tension between capacities and balance is played out in planning practice, though the themes are familiar in the context of housing, transport and many other forms of development. Berkshire's replacement Minerals Local Plan is a particularly interesting case because it draws explicitly on concepts of sustainability. A key argument in the 1993 draft plan was that production of aggregates (in this case, sand and gravel) at the rates indicated in South East Regional Guidance after 1996 would breach the environmental capacity of the County.[10] This conclusion had been reached after taking account of national and local environmental constraints, assessing the overall impacts of mineral working and engaging in two rounds of public consultation. The plan proposed to reduce annual production (2.5 million tonnes per year) by 3 per cent per annum after 1996. Strong objections were lodged by mineral operators and their trade associations on the grounds (*inter alia*) that environmental considerations had to be balanced with the need for local land won aggregates and that the County was giving insufficient consideration to the latter.

At the ensuing public inquiry the process of selecting sites in which minerals extraction could be contemplated, which was closely aligned to the County's judgment of environmental capacity, became an important issue. The Inspector was persuaded that this process had been 'thorough and correct' but he was not convinced that the constraints in excluded areas were always too severe to be overcome or that they justified imposing limits on development for the period of the plan. In effect, he questioned the County's ability to assess 'environmental capacity': indeed he did not accept the use of the term as a legitimate planning concept. Particularly revealing is the concern expressed in the Inspector's report abut the implications of a more widespread adoption of Berkshire's capacity-constrained approach: if other counties followed suit, he argued, 'severe constraints' might be placed on the production of aggregates which had a 'vital role to play in the national economy'.[11] He recommended reversion to a policy in line with the Regional Guidelines.

The Berkshire minerals experience illustrates two important difficulties for planning authorities and others who seek to interpret sustainable development in terms of environmental capacity. For one thing, they have to be able to give meaning to the term in ways which command agreement and respect, and to show that particular forms of development will necessarily breach capacities. This is problematic, because deciding what is critical in the environment, and what constitutes 'quality' to be maintained, clearly involves judgment, an issue which is discussed in more detail below. Secondly, they face a still powerful orthodox view that social and economic 'needs' must be satisfied: hence the Inspector's concern that capacity constraints, aggregated over counties and regions, might threaten the imperative to meet the 'national need' for minerals. The 'balance' so often invoked is already weighted: it tends to mean balance on a case-by-case basis, within a general framework in which it is assumed that certain needs should be met. This brings us to another tension starkly exposed by rival claims to the high ground of sustainable development: the tension between meeting and managing consumer demands.

Meeting or Managing Demands?

Society's demands for a wide range of goods and services are growing. In many cases—roads are a pertinent example—making provision in an attempt to meet demand has already became a source of bitter conflict. Recognising that the necessary facilities require the consumption of valued environments, the rhetoric of sustainable development subtly places emphasis on 'needs' rather than demands and on the modification, moderation and perhaps even reduction of the latter. In principle, this philosophy appears to be widely accepted; in practice, serious political commitment to demand management is barely detectable. This ambivalence is indicative of a deep tension between the assumption that consumer demands should be met as far as possible and the view that they should be moderated in the light of environmental constraints. This tension is linked, of course, to the conflict between capacities and balance, since the case for demand management depends essentially on the argument that environmental capacity, at some level, is limited. The planning system, required on the one hand to make provision to meet demands and on the other to be an instrument of environmental protection, is unavoidably implicated in this tension.

The issues are most sharply exposed in those sectors (such as minerals and housing) where projections of demand are produced by the state and where local planning authorities are required to make appropriate land allocations in development plans. Projections are usually based on a positive correlation of demand for some good or service with GDP (or, in the case of housing, on demographic change and anticipated household formation). The orthodox view is that they represent a best estimate of future legitimate needs for 'essential' goods and services, and that the land use planning system has a key role in ensuring that these needs are met at the lowest practicable environmental and social cost. It is possible, of course, to challenge the *accuracy* of forecasts based on projections and many have tried to do so, though with rather limited success. More significant is a challenge to the presumption that demands for goods and services must be met and to the role of public policy (including the planning system) in achieving, or facilitating, this objective. This second challenge has been given powerful new impetus by the rhetoric of sustainable development and raises important questions about the definition, articulation and status of different demands and their relationship with social needs.

Once more the issues can be illustrated with reference to minerals extraction, though again they are familiar in relation to many other development pressures. In line with the new emphasis on sustainable development, minerals planning policies now stress anticipation and mitigation of impacts, efficiency of use and increased use of secondary or recycled materials, while avoiding difficult questions about the nature or legitimacy of demands. The Government has also stressed (as it does in the case of traffic and housing) that projections are not 'targets', and there is an acknowledged need to 'test' regional allocations in the planning process. It is not clear, however, what the outcome should be if the 'testing' reveals (as Berkshire claimed) that meeting demand would have unacceptable environmental consequences. Frequent re-assertion of the claim that an 'adequate and steady supply' of minerals is 'essential' reveals an underlying assumption that planning will somehow be able to achieve the 'best balance' between conflicting needs. Indeed, new draft guidance on planning principles is quite explicit about the potential for 'spatial fixes': the planning system should promote sustainable development by 'helping to provide for necessary development *in locations* which do not compromise the ability of future generations to meet their needs'.[12] However, conflict over a wide range of developments, repeated in many locations, suggests that the scope for spatial fixes is diminishing and that more troublesome questions about consumer demands will have to be confronted.

The logic of stronger interpretations of sustainable development leads inexorably (and for some, unexpectedly) in this direction. To argue that development must respect environmental capacities is tantamount to saying that certain environmental 'needs' are essential too. The deep dilemma, evaded in policy statements but regularly confronted on the ground, is what to do when the scope for spatial and technical fixes, at least for any meaningful planning period, is exhausted and the need for vital environments is incompatible with the need for 'essential' goods and services. The answer must be (as we are beginning to see in the case of transport) a more careful scrutiny of the nature and legitimacy of demands—for material as well as environmental goods—and their relation to the concept of 'need'. This, of course, is the root of the tension, for while 'demand management' sounds innocuous enough, distinguishing needs from demands raises fundamentally difficult questions.

There is no space here to do justice to these issues. Suffice it to note that demands for environmental quality have to be justified in planning policies and decisions with some rigour, especially when they relate to assets which are not nationally designated. At the Berkshire inquiry, the Inspector accepted that the Minerals Plan offered an opportunity to 'test' the post-1996 aggregates production figures, but insisted that the County would have to be able 'to demonstrate very clearly the reasons why it cannot maintain a production of 2.5 million tonnes per year . . .'.[13] The industry, in contrast, was not required to 'demonstrate very clearly' why it was essential that such a level of production be maintained. The planning system does not normally require applicants to demonstrate the need for their proposed developments: it has been sufficient to cite projected demands, market forces or the (alleged) interests of the local or national economy. In this framework, to question 'need' for goods and services is to challenge growth, prosperity and consumer sovereignty.

One further point should be made here. An inevitable result of the unassailable status of 'need' in many sectors is to cast objectors, and local planning authorities, in the role of NIMBYs. Resistance to development in their own areas must mean (if total demand is to be satisfied) that environmentally damaging projects are shifted elsewhere: thus those who resist development can be represented as narrowly self-interested and can be more readily overridden. Where development *is* successfully resisted it may be displaced to remote or disadvantaged locations; indeed, as noted above, 'spatial fixes' are to some extent institutionalised in the planning system as a legitimate means of achieving the 'best balance' of social, economic and environmental objectives. But they give rise to a new set of problems, because spatial variations in the impacts of different activities are mapped onto widely varying social and economic circumstances. Environmental effects are multi-dimensional, involving local impacts on amenity and economic interests as well as less tangible—but no less important—changes to the scientific, existence and intrinsic values of environments: it is often far from straightforward to ascertain that activity in one location will have 'less' impact than in another. If the shift is to a remote area (as is the case with the new emphasis on coastal superquarries in Scotland), fewer people may be directly affected, but almost by definition, this is at the expense of relatively wild and unspoilt places.

There is further complexity. Remote or economically peripheral communities often welcome any economic activity as a source of jobs and local income: there may be many different local interpretations of sustainable development. But environments are not necessarily 'owned' (except in the most literal sense) by local people, or even by present generations: they have value to others and to the nation as a whole, sometimes, though not always, reflected in national designations. In defining what is sustainable, new questions about 'environmental subsidiarity' are superimposed on the old conflict between national strategy and local democracy.

'But that's subjective': Facts and Judgments in the Planning Process

it will be apparent by now that underlying all of these tensions are important questions about the status of the information and judgments on which planning policies and decisions must be based. How should environmental capacities be determined, by whom and at what geographical scale? On what basis can fragile environmental values—beauty, tranquillity, the intrinsic worth of non-human or cultural assets—be compared with consumer preferences expressed in markets? How do we decide if such preferences merit treatment as genuine social needs, and does it fall within the remit of planning to make this distinction?

Some inputs to the planning process are 'objective' in the sense that they are empirically verifiable. Others, such as judgments about beauty of landscape or nuisance caused by noise, are clearly more subjective, though they may be widely shared. Once we move beyond claims which most people would accept as 'factual' towards those which most would recognise as involving judgment, the territory becomes less certain. Nevertheless, information claimed to be objective is generally accorded higher status than that which involves subjective value judgments. At the Berkshire inquiry, for example, the 'subjective' delineation of areas in which minerals exploitation would or would not be considered was a major target for objectors, while the County's share of aggregates production (derived ultimately from demand projections) acquired a relatively solid status. On the Isle of Harris where a superquarry is proposed, objectors' submissions about the image and 'existence value' of the island as a place of peace and tranquillity were dismissed by supporters of the quarry as 'hypothetically metaphysical'.[14]

This dichotomy of hard and soft information—highlighted by the need to determine what is 'sustainable'—can be called into question for at least two reasons. First, it is not at all clear that the categories are consistent or correct. A planning authority's judgment that local landscapes or habitats constitute critical environmental capital, for example, is no more subjective than the indiscriminate assertion that some product or service is 'essential': after all, consumer demands are themselves expressions of subjective preferences, and the view that they should be met is a value judgment. Second, the relegation of what is regarded as subjective implies that it involves merely an expression of preference or taste which can make no claims upon the agreement of others. This is unfortunate, because the planning process affords one of the most important opportunities to make defensible *intersubjective* judgments of value, applying appropriate standards of argument, criticism and debate. Arrived at in this way, judgments about environmental value (and capacity) provide no less valid a basis for policy than demand projections or other inputs traditionally afforded a more solid status. Currently, of course, planning is a far from perfect forum for intersubjective judgment: the need for rigour too readily provides an excuse for spurious quantification, and identifying proper roles and mechanisms for expert, political and public input to the process remains a significant challenge. What is interesting, however, is that attempts to interpret sustainability in planning practice underline the need for defensible judgment as a legitimate part of the political process. The ideal of the planning system as a forum for such judgment—as an exercise in civic republicanism rather than a means of organising consumer preferences—should certainly not be abandoned.

Conclusions

This brief review of planning practice reveals a number of unresolved questions lurking beneath the surface of sustainable development. This is not surprising in itself: what is interesting is that contradictions which can be glossed over at the level of principle are exposed when the need to make policies and decisions for real places demands the concrete interpretation of abstract ideas. In these circumstances, the consensual rhetoric of sustainability—'living within our environmental means', for example, or 'moderating and managing consumer demands' soon gives way to conflicting philosophies and values. An important conclusion, therefore, is that while sustainable development provides a new framework for the articulation of divergent positions, it fails to reconcile them: indeed, when differences are profound, they become internalised, and perpetuated, in conflicting inter-pretations of what it means for development to be sustainable.

For this reason sustainability is interpreted *through* the planning process—in policies and guidance, plans and decisions—in various ways. Plans which stress the management of demand within environmental capacity constraints claim to be pursuing sustainable development, as do policies seeking to balance environmental objectives with other social needs and priorities. These claims are contested in comment and debate, in Examinations in Public and at planning inquiries. Those who argue for a strong interpretation of sustainable development are called upon (explicitly or implicitly) to show how environ-mental capacities can be defined, while those who are willing to trade off environmental and economic objectives must explain how, and at what level, 'balance' can be achieved, especially if some goods and services are deemed in advance to be 'essential'. A second conclusion, therefore, is that interpret-ing sustainable development—to the surprise of some of its advocates—prompts scrutiny of assumptions about 'needs' and 'demands' and requires consideration of which goods, services and environmental assets we value most. The question of whether planners should simply respond to demands or seek to moderate them is one that refuses to go away. The planning system could provide a forum in which to differentiate between genuine social needs which can *only* be met by expanding supply (of minerals, energy, housing or whatever), needs which might be satisfied in other ways, and demands which might reasonably be regarded as subordinate to important environmental considerations. It currently allows for discussion of these issues by default rather than by design.

Perhaps the greatest challenge for planning is to find ways of integrating different perspectives which, in their extreme form, might be mutually exclusive. Most people—even supporters of 'strong' sustainability—would agree that few things in a democratic society can be deemed inviolate in all circumstances. This does not mean, however, that everything is negotiable. The important questions are not whether we respect capacities in some absolute sense or agree always to trade off conflicting objectives. Rather, we should be asking which characteristics of the environment merit protection for present and future people or for their own sake, and what is appropriate to put into (or keep out of) the balance at different levels—local, regional, national and international. Resolving these questions cannot be reduced to an exercise in technical rationality (through the various forms of 'evaluation'), nor can it be left to markets, which systematically undervalue environmental resources. Science helps, but cannot tell us how we wish to live. So we reach the third conclusion—that planning for sustainable development requires rigorous, intersubjective judgment. The procedures for arriving at such judgments, and their treatment within the decision-making process, would benefit from substantial re-assessment.

Finally, if sustainability is to be defined on the ground in terms of environmental capacities, this means formulating policies concerning the amount and type of development that will be tolerable in particular geographical areas over given time periods. Even the House of Lords Select Committee, though remaining sceptical about the concept of capacities in general, felt that there must be limits to the capacity of the environment to host particular forms of development at any one time. If this is so, it is hardly surprising that translating concepts of sustainability into planning practice has proved much more difficult than expected, for it shows that there are circumstances in which sustainable development reduces, after all, to the older discourse of limits to growth.

Biographical Note

Susan Owens is a Lecturer in Geography at the University of Cambridge and a Fellow of Newnham College. She has long-standing research interests in environmental issues and policies, and has written widely on interpretations of sustainable development in the land use planning process.

Notes

1 Secretary of State for the Environment *et al.*, *Sustainable Development: the UK Strategy*, London, HMSO, 1994, p. 221. For a more detailed discussion of land use planning and sustainable development see S. Owens, 'Land, limits and sustainability: a conceptual framework and some dilemmas for the planning system', *Transactions of the Institute of British Geographers NS*, 1994, Vol. 19, No. 4, pp. 439–56; also S. Owens and R. Cowell, *Rocks and Hard Places: Mineral Resource Planning and Sustainability*, London, Council for the Protection of Rural England.
2 World Commission on Environment and Development, *Our Common Future*, Oxford, Oxford University Press, 1987.
3 House of Lords Select Committee on Sustainable Development, *Report, Volume I*, London, HMSO, 1995, p. 10.
4 One of the most accessible statements is in D. W. Pearce, A. Markandya and E. Barbier, *Blueprint for a Green Economy*, London, Earthscan, 1989. For an analysis more clearly relating to planning, see M. Jacobs, *Sense and Sustainability*, London, CPRE, 1993 and for an overview see S. Owens, *op. cit.*.
5 See, for example, W. Beckerman, *Small is Stupid*, London, Duckworth, 1994; S. Grigson, *The Limits of Environmental Capacity*, Report to the House Builders' Federation, London, Barton Willmore Partnership, 1995; House of Lords Select Committee on Sustainable Development, *op. cit.*, paras 2.1, 3.18, 4.8, 5.7; R. Solow, *An Almost Practical Step Towards Sustainability*, Fortieth Anniversary Lecture, Resources for the Future (RFF), Washington DC, RFF, 1992.
6 The potential to achieve the same (or more ambitious) ends with less environmental impact—to reduce the 'environmental co-efficient of economic growth'—is

disputed. Some see it as almost limitless; others take a more precautionary attitude to the potential to predict and mitigate impacts and to the feasibility, and acceptability, of replacing environmental assets that are damaged or lost. These divergent attitudes constitute a significant tension in themselves but it is not possible to elaborate on these issues here.

7 Secretary of State for the Environment *et al.*, *op. cit.* (note 1), para 35.4, emphasis added.

8 Secretary of State for the Environment *et al.*, *Response to the House of Lords Select Committee on Sustainable Development*, London, HMSO, 1995, paras 7 and 8. See also Secretary of State for the Environment *et al.*, *Biodiversity: The UK Action Plan*, Cm 2428, London, HMSO, 1994.

9 Department of the Environment *Planning Policy Guidance: Nature Conservation* (PPG 9), London, Department of the Environment, 1994, para A15. (This originates in European rather than national legislation. It also requires environmental compensation when damage is unavoidable.) A similar philosophy could be said to apply to 'critical cultural capital': important listed buildings must not be demolished 'simply because redevelopment is economically more attractive'. (Department of the Environment and the Department of National Heritage 1994, *Planning and the Historic Environment* (PPG 15), London, Department of the Environment, para 3.17.)

10 Royal County of Berkshire, *Draft Replacement Minerals Local Plan for Berkshire*, Reading, 1993. Projections of national demand for aggregates are disaggregated by region and each County is then given an apportionment for which provision has to be made. A fuller account of this process, and of the Berkshire case, may be found in Owens and Cowell, *op. cit.*.

11 M. J. Brundell, *Inspector's Report of the Inquiry into the Replacement Minerals Local Plan for Berkshire, 21 September to 16 November 1993*, Bristol, Planning Inspectorate, 1994, para 3.2.7.

12 Department of the Environment, *Planning Policy Guidance: General Policy and Principles* (PPG1), Consultation Draft, London, Department of the Environment, 1996.

13 M. J. Brundell, *op. cit.*, emphasis added.

14 Martin and Abercrombie, closing submissions to the Inquiry (Autumn 1994 to Spring 1995) into a proposal by Redland Aggregates for a superquarry at Lingerbay, 1995, p. 53. At the time of writing a decision on this case is still awaited. For details, see Owens and Cowell, *op. cit.*

The Political Economy of Environmental Tax Reform

STEPHEN TINDALE

IN recent years British politics has been transfixed by tax. Will Tories or Labour cost you more? This debate is important, but it is only part of the story. Equally important is the way the money is raised. 'How' matters, as well as 'how much'.

Across the EU, roughly 50 per cent of all taxes are levied directly or indirectly on labour (mainly through income tax and social security contributions), and less than 10 per cent on natural resources. The burden of taxation on labour has increased steadily, from around 30 per cent in 1960, while that on natural resources has actually declined. Britain conforms to this picture.

'Environmental tax reform' is the rather ungainly name given to the proposal to shift the burden of taxation off 'good things' which society wants to encourage, like employment, income, savings and value added, and onto 'bad things' we want to discourage like pollution, resource use and waste. Proponents of environmental tax reform argue that it would not only help the environment, but could also generate substantial numbers of new jobs—this is referred to as the 'double dividend'.

This sounds like common sense. But not everyone agrees, which is why only the Scandinavian countries and The Netherlands have so far started down the road towards environmental tax reform. This chapter considers first the arguments in favour of tax reform and the potential benefits, and the criticisms which are most commonly made of the proposals. It then considers the key issue of who would win and who would lose, leading to a consideration of the political economy of the debate and the type of proposals which might make headway.

Arguments for Environmental Tax Reform

Assuming that we need to do more to protect the environment, economic theory suggests that taxation is often the most efficient approach, encouraging those companies which can alter their behaviour most easily and cheaply to do so, while those for whom adjustment would be expensive simply pay the tax. Pollution will thus be reduced at least cost to society. A regulatory approach, which required all companies to alter behaviour equally, would be less efficient.

This argument is now widely accepted. The European Union's Fifth Action Programme on the Environment, for example, promises to make much wider

Published by Blackwell Publishers, 108 Cowley Road, Oxford OX4 1JF, UK and 350 Main Street, Malden, MA 02148, USA

use of 'market mechanisms'. Two caveats need to be entered. First, taxation is not appropriate for every environmental issue: nuclear power stations should be closed down, not taxed. Secondly, taxation will often work best in combination with regulations or other government policies. For example, higher petrol prices will be most effective if accompanied by good public transport, minimum efficiency standards for vehicles and sensible land-use planning policies. Nevertheless, in general the argument that taxation is more efficient than regulation is a valid one.

A second argument for environmental tax reform is that green taxes are an efficient way for governments to raise the money which they need to pay for welfare payments, law and order, health, education and so on. Most taxes are distortionary to a greater or lesser extent since they change relative prices. While taxes will often reallocate or redistribute resources in a desirable way, there is often some deadweight loss involved.

However, market prices do not always reflect the full cost of an activity. In this case taxes can enhance efficiency. Some of the costs of pollution, for example, are not borne by the producer but dispersed on to society. The producer therefore does not need to charge for them. So unlike most taxes, which damage economic efficiency, environmental taxes can actually improve it by making prices tell the full story. Kenneth Clarke accepted this argument in his 1994 Budget speech: 'in some cases, taxes actually do some good, by helping markets work better and by discouraging harmful or wasteful activity.' Green taxes should be used to raise revenue currently raised by taxes which are distorting, like those on employment.

What Could Environmental Tax Reform Achieve?

To give an indication of what tax reform could deliver, in terms of environmental improvement and employment increases, IPPR commissioned Cambridge Econometrics to model a package of different measures for the period 1997–2005. The main elements of the package are:

- A commercial and industrial energy tax
- A waste disposal tax
- A higher road fuel escalator
- A quarrying tax
- An office parking tax
- An end to company car tax perks

The revenue so raised is to reduce employers' NICs (National Insurance Contributions). According to the model, the reform would deliver substantial reductions in emissions of key pollutants, including a reduction of 9.5 per cent in total carbon emissions in 2005. The main reductions are from road transport (19 per cent), iron and steel (18 per cent), chemicals (12 per cent) and other industry (10 per cent). There is also a 16 per cent reduction in the amount of waste produced.

The main economic impact, as one would expect, is that production has become more job-rich. There are 252,000 extra jobs in 2000, and 717,000 in 2005. Unemployment falls by 300,000; this difference is because not all those who fill the new jobs are registered as unemployed. Two-thirds of the new jobs are full-time. All regions gain jobs, though some do better than others. The most significant increases are in health and social work (141,000), business services (91,000), education (61,000) and construction (50,000), but employment in manufacturing industry is also simulated to increase substantially.

The impact on other macroeconomic variables is small—again as one would expect with a package which leaves the overall level of taxation unchanged. GDP is virtually unchanged—up 0.31 per cent in 2000, down 0.03 per cent in 2005, both infinitesimal in relation to error margins. There is an increase in inflation of less than 0.25 per cent a year, again insignificant in relation to model error. The balance of payments deteriorates very slightly, since the change diverts activity from traded sectors to sectors whose output is traded less or not at all.

Economic modelling is necessarily an inexact science. The results should not, therefore, be taken as a quantitatively-reliable guide to what will happen. What matters is the broad tendencies which they reveal—negligible impact on GDP and inflation, a significant increase in employment—rather than the precise figures.

IPPR's results reveal similar trends to studies of environmental tax reform carried out elsewhere in Europe. For example, the Deutsches Institut für Wirtschaftung carried out a study into the impact of a unilateral tax reform in Germany, involving progressively rising energy taxes and reductions in employers' social security contributions and a per capita 'eco-bonus'. The findings were a predicted increase in employment of 600,000 after 10 years, and insignificant impacts on economic growth and competitiveness. Estimated energy use declined by 7 per cent.

The impact of a broader tax reform package, including a carbon/energy tax, higher transport taxes, agricultural input taxes, those on water and pollution together with a range of non-fiscal measures, has been modelled for the European Commission by a consortium of consultants led by DRI. They too predict that if employers' social security contributions were reduced, GDP would be 1 per cent higher than in the reference case. Employment is predicted to be 2.2 million higher by 2010, reducing the European unemployment rate by 1 per cent.

Environmental Tax Reform in Practice

The double dividend hypothesis lies behind a series of tax reforms carried out by Northern European countries since 1990. Sweden introduced its tax reform in 1990 and 1991, resulting in a major redistribution of the tax burden (representing 6 per cent of GDP). The main concern was to reduce the very high rates of personal income tax. On the environmental side there were a number of changes: new taxes on sulphur and nitrogen emissions were levied, energy was brought within the VAT system, and a carbon dioxide tax was introduced. Energy-intensive industries were given transitional tax relief until 1995.

The main impact so far has been on fuel switching, especially in the district heating sector where there has been a switch to biofuels. According to the Environment Ministry, the environmental impacts of the package have exceeded their expectations. However, the European Environment Agency argues that the impact of the energy taxes is hard to evaluate due to the short period of operation, and that sulphur emissions have dropped by only 6 per cent as a result of the tax. The Swedish Government is in the middle of an assessment of the economic impacts.

The tax reform in Norway also took place over a two-year period. In 1991, a range of environmental taxes was introduced, including sulphur and carbon taxes. In 1992 the rates of personal and company taxation were lowered. The carbon tax led to reductions of 3–4 per cent in carbon emissions between 1991–93, from a rising trend.

In 1994 a major tax reform package in Denmark reduced income taxes and increased environmental taxes on households. In 1996 the 'Danish Energy Package' extended CO_2 and energy taxes to industry and introduced a sulphur tax. The most substantial tax is on energy used by industry for heating purposes. Companies which agree to make large investments to improve energy efficiency receive a partial rebate. The Government estimates that the new taxes will reduce Danish CO_2 emissions by around 5 per cent. Revenue is recycled primarily through reduced employers' social security contributions, though some is available for investment incentives, and some will also be targeted at the small business sector.

In January 1996 The Netherlands introduced a carbon/energy tax on small energy users: households and small commercial establishments. The Government argues that these are sectors which it is difficult to reach with other policy measures such as permits or long-term agreements, but admits also that larger users have been excluded to avoid possible economic damage. The tax is closely modelled on the proposed EU tax, and will eventually reach the same rate of $10 per barrel of oil equivalent. It applies only for use between 800–170,000 cubic metres of gas and 800–50,000 kWh of electricity. The lower limit is in recognition of the fact that it is not possible to reduce gas or electricity use to zero. The revenue from the new tax will be used to reduce direct taxes.

The Debate in the UK

The concept of a green tax reform is not as widely debated in the UK as in many other European countries, but it has been gaining support gradually in recent years. Conservative Chancellor Kenneth Clarke signed up to the principle when he linked the new landfill tax to a reduction in employers'

National Insurance Contributions, saying: 'I want to increase the tax on polluters, and make further cuts in the tax on jobs.' However, the landfill tax raises a comparatively small amount of money, around half a million pounds a year, so the shift cannot be said to be significant, and the 1995 and 1996 Budgets contained no new taxes on polluters.

A number of Government Advisory bodies have supported the principle of environmental tax reform. The Panel on Sustainable Development, set up by John Major after the Rio Summit to advise him on environmental matters, has called for 'wider use of economic instruments, and a gradual move away from taxes on labour, income, profits and capital towards taxes on pollution and the use of resources, including energy'. The Advisory Committee on Business and the Environment (ACBE), which includes representatives from major energy, construction, retailing and financial services companies, has stated that 'ACBE welcomed the Chancellor's statement in his 1994 Budget that in future the Government would be looking to shift the burden of taxation from wealth creation to resource use and pollution'.

Environmental groups have mostly overcome earlier scepticism about green taxes and many, including Friends of the Earth and the RSPB, now campaign in favour of green tax reform. The recently launched Real World coalition, which brings together environmental and social policy groups (itself a significant development), strongly supports a tax shift.

Although the Left has traditionally resisted the notion that the cost of labour contributes to unemployment, it is becoming increasingly widely accepted that labour costs should be reduced in order to encourage employers to create more jobs. For example, the report of the Commission on Social Justice calls for:

The development of a tax and benefits system which provides incentives, not disincentives, to employment. This will require . . . gradual reduction in taxes on employment, particularly for less-skilled and lower-paid jobs.

The Labour Party, in its 1994 Environment Policy Statement *In Trust for Tomorrow*, committed itself to:

a long term, gradual change to the way in which the economy is organised, to ensure that it encourages 'goods' such as employment, value added, investment and savings, and discourages 'bads' such as pollution and resource depletion.

The Liberal Democrats have committed themselves to a phased introduction of a carbon tax on energy production, with safeguards for domestic consumers. Renewables would be exempted. Revenues would be recycled to reduce other taxes, though these are not specified.

Overall, the debate in the UK is slowly widening, and there are some influential voices supporting reform. However, it is much easier to sign up to a principle than to support and implement specific reform packages. Once specific measures are proposed, support often evaporates. What, then, are the forces ranged against environmental tax reform?

Industrial Lobbies

The failure of the European Commission's carbon/energy tax proposal was partly due to the British Government's opposition on the grounds that the EU should not be involved in setting taxes, but partly also due to strong lobbying from industrial interests. The Commission proposal included safe-guards to protect Europe's competitiveness: major energy users would be exempted, border tax adjustments would be made on imports, and the whole package would be conditional on similar action by the main competitors. Moreover, member states would be encouraged to return the revenues to business in the form of lower social security contributions. Yet still the energy intensive industries, and the energy sector itself, lobbied hard against the proposal, opposing the very concept of increasing the cost of industrial inputs.

The power of industrial lobbying is demonstrated also by Scandinavian experience. The Swedish Government argued in 1991 that it expected the EU to introduce its carbon/energy tax. When it became clear that this was unlikely, Swedish industry was able to pressure the Government to reduce the tax rate paid by companies, and the lost revenue was recouped by raising the rate paid by households. But in 1996 the rate of tax on industry was doubled, even though the EU had still not acted. In Norway too the package had to be modified after it was introduced, to take account of the impact on energy industries, and to reflect the fact that the EU had not introduced energy taxes as expected.

Not all industry is opposed to environmental tax reform, however. A number of leading German companies, including domestic appliance manu-facturer AEG, have come out in support of green tax reform, running full page advertisements demanding that politicians 'put this long-overdue tax reform into practice'. Those concerned to promote tax reform need to build coalitions with potential 'winner' sectors, and isolate the losers.

The most obvious fault line in industry is between resource-intensive sectors and labour-intensive ones. A tax shift which reduces the price of labour while increasing the price of energy, for example, will benefit construction, health and education, financial services and retail and disad-vantage chemicals, iron and steel, non-ferrous metals and non-metallic minerals. One can expect the latter to mobilise strongly against proposed changes, and this was indeed the experience with the carbon/energy tax. Sectors which would have gained took little interest. It is axiomatic that in any proposed tax reform the losers will shout louder than the winners.

However, not all energy-intensive sectors will be equally disadvantaged. Assuming that the tax shift is carried out by one country acting alone, sectors which are highly traded, such as chemicals or non-ferrous metals, may face a competitive disadvantage, at least in the short term. Sectors like non-metallic minerals or (in the UK context at least) iron and steel, which are less highly traded, ought to be less concerned about this effect. This does not mean that such sectors will not oppose the proposals, but it may make their opposition less deep seated.

Is concern about competitiveness justified? Sudden changes in the cost of energy in any one country will have an impact on its ability to compete. The impact of gradual increases, leaving industry time to adapt, may however by positive. Countries with high energy prices, such as Germany and Japan, tend to be more competitive and wealthier than low-cost nations. While this correlation is not conclusive and does not imply simple causation, it suggests at the very least that there is no necessary connection between low energy prices and economic success.

There are also winners and losers in a different sense. Some industrial sectors do well out of existing, unsustainable patterns of economic activity—society's profligacy is, to them, a source of increased profits. Oil companies, energy generators, aggregates producers and waste disposal companies fall into this category. The sectors which would do well out of a different pattern of production—energy efficiency industries, clean technology manufacturers and so on—are by definition less successful at the moment.

The fossil fuel and chemical industries are alert to the dangers to their interests posed by environmental concerns; their lobbying operations are well-resourced, well-organised and not overly-principled. This is what John Gummer was referring to in his speech to the Climate Change Conference in July 1996 when he attacked 'the purveyors of falsehood who put their selfish concerns before the interests of the world community' and counselled that 'none of us should give way to the commercial propositions which are hidden by the pseudo-science of those who pretend that what the world knows to be true can be put on one side because of an individual's desire to promote his particular and prejudiced view'. When a Conservative Cabinet Minister talks about big business in this way, one can safely assume that there is some serious special pleading going on behind the scenes.

Those lobbying for change tend to be far less powerful than those supporting business as usual. Many of the firms they wish to promote are either small or not yet in existence, so are not very helpful in fundraising terms. Britain now has an Environmental Industries Commission, which does useful work. But it cannot compete with the big corporations in the all-important matter of wining, dining and duping politicians.

However, proponents of change have a powerful and rich new ally: the insurance sector. Insurance companies are losers in a very literal sense when environmental policy fails to prevent damage: they have to foot the bill. The damage wrought by storms, floods, hurricanes and so on is spiralling upward. In the first three years of the 1990s, there were twice as many major windstorms worldwide as in the whole of the 1980s, causing damage worth $20 billion. This is why the insurance industry is taking global warming so seriously. The major companies all have global warming units, and lined up with environmentalists in pushing for greater action at the latest session of international negotiations on the Climate Change Convention.

The 'Poverty Lobby'

The most commonly heard reaction to proposals for environmental tax reform in the UK context is that green taxes penalise the poor. Tax reform which secured environmental or economic improvements at the expense of low-income groups would rightly be rejected by most people. Suspicion of higher environmental taxes can be expected from the coalition of groups working on behalf of the poor, the elderly, the disabled and the otherwise vulnerable, sometimes referred to as the poverty lobby (the term is used here with no pejorative overtones).

The best way to counter the claim that environmental taxation is regressive is to break it down into its constituent parts. First, there are different ways of looking at progressivity and regressivity. One can look simply at the incidence of taxation: an income tax can be more progressive than a sales tax. Or one can look at what the government proposes to do with the revenues. Income tax used to subsidise opera tickets is *less* progressive than a sales tax used to pay welfare benefits. A third approach is to compare the impact of action with the impact of inaction. Environmental degradation often impacts most heavily on poor communities. Inner-city residents suffer more from transport-related pollution, and in the UK this generally means poorer people. Working class communities have less bargaining power in lobbying against undesirable environmental developments being located in their 'backyards'. And poor people will be less able to protect themselves from the health effects of environmental decline. Environmental issues affect everyone, but they do not affect everyone equally. The blanket assertion that environmental taxation is regressive greatly oversimplifies the picture. Cuts in government expenditure or a failure to act to protect the environment may be even more regressive.

Secondly, we need to ask: which particular environmental taxes are being considered? There are legitimate concerns about the impact of some green taxes, particularly the taxation of domestic energy. This is essentially a British problem: in other countries the superior state of the housing stock means that fuel poverty is almost unknown, and spending on fuel correlates much more closely to income. This is why all other EU countries levy VAT on domestic fuel, and why Scandinavian countries (with colder climates than Britain) were able to impose new taxes on the domestic sector without provoking political protest.

In Britain, taxation of domestic energy is highly regressive. The poorest fifth of households spend 12 per cent of their budget on fuel; the richest fifth just 4 per cent. The poor are less able to cut back on fuel use by changing equipment or installing energy efficiency measures, which can have a high capital cost. And those who live in private rented accommodation have less incentive to invest in insulation since they may not stay long enough to recoup the cost, while the landlords have little incentive because they do not pay the bills. This suggests that tax might not be the appropriate way to encourage efficiency in the domestic sector.

Other green taxes, such as higher petrol taxes, are less problematic. Petrol taxation is broadly progressive over the population as a whole, since very poor people do not own cars—although they may aspire to own them. (However, it is regressive within the car-owing community, impacting more on poor car owners than rich car owners, and on the rural poor most of all. Some revenue could be used to compensate them, or offsetting tax reductions could be targeted.) Green taxes levied on industry would be no more regressive than any other direct or indirect tax which increases the cost of production. Opposition to these changes is likely to come not so much from the poverty lobby as from consumer groups.

The Consumers Lobby

Unsustainable consumption is at the root of environmental problems. Industry produces, and therefore pollutes, in order to satisfy consumer demand (although it is true that some of this demand is stimulated by industry's own efforts). Yet consumers' organisations tend to assume that those they represent are innocent parties, vulnerable to misleading claims about 'environmental friendliness', but otherwise not much involved in the environmental debate. An example of this is a paper from the Consumers in Europe groups which argues that a carbon tax should be rejected because it 'would not be applied close enough to the source of pollution to encourage energy efficiency directly and extra costs incurred by manufacturers are likely to be passed straight on to the consumers'. But it is in fact desirable that the cost to consumers of energy-intensive or other environmentally-damaging goods should increase.

Consumer groups can be expected to support anything which increases the provision of information, such as the EU's eco-labelling scheme or energy labels. Some groups have also been prepared to support mandatory energy efficiency standards. But most have opposed, and can be expected to continue to oppose, environmental taxes. This is somewhat illogical. Mandatory efficiency standards will have the effect of taking some of the cheaper, less efficient models off the market, and may increase the cost of others. The choice between regulation and taxation may not, in fact, be so great.

Trade Unions

Environmental tax reform would lead to higher overall levels of employment. At one level one would therefore expect trade unions to support it. However, the situation is not so straightforward. Trade unions have historically campaigned to protect existing jobs, rather than supporting reforms which might lead to higher levels of employment overall, but would involve job losses in some sectors. However, there are signs of change in the trade union attitude. For example, MSF has been campaigning for many years for government policies to promote clean technologies, recognising that this could create thousands of jobs, and arguing that 'environmental diversification' should be an essential part of any industrial strategy. UNISON has sponsored research into the employment-creation potential of environmental policy. And John Edmonds, General Secretary of the GMB, has endorsed the principle of environmental tax reform.

Concern has been expressed by some trade unionists about the weakening of the contributory principle if employers' National Insurance Contributions are reduced. The contributory principle is a fiction—NICs are not nearly large enough to cover what they are supposed to fund—but it may nevertheless be important.

Building a Coalition for Environmental Tax Reform

It is clear from the above analysis that there will be strong opposition to any proposals to change the burden of taxation, and the potential losers will mobilise and lobby harder than the winners. Nevertheless, it should be possible to design a package which maximises political support and under-cuts opposition. This can be done in two ways: by selecting the proposed tax changes carefully, and by involving key groups in drawing up the proposals.

In the light of recent political experience, the main potential opposition to reform in the UK is from those concerned about fuel poverty. As we have seen, domestic energy taxation in the UK is highly regressive. However, tax measures could be made progressive if the revenue were used to give a lump sum payment to each individual, or possibly to each household. This would be substantially more than the extra tax paid by poorer households, but much less than the extra tax paid by the rich. Thus the package overall becomes progressive. Alternatively, one could exclude domestic energy taxation altogether from the reform package. This was the approach used in IPPR's work on green taxes. This would effectively avoid opposition from the poverty lobby.

Trade union opposition could be avoided if tax reductions were used to reduce something other than NI contributions. Opposition to reductions in employers' NICs should not be taken as opposition to the concept of green tax reform, only as opposition to one variant of it. The revenues could equally be used to cut income tax, business rates, corporation tax or VAT. However, economic modelling suggests that cutting NICs would be the most effective at creating jobs, so the need to build political support will have to be balanced against the desire to secure economic goals.

Consumer groups could perhaps be mollified if some of the revenue was used to reduce VAT. In addition, the fact that labour costs would be reduced as energy costs increased would mean that overall changes in the RPI would be very small. Any residual opposition from consumers would need to be met with robust arguments about what the 'polluter pays principle' actually means—consumers may not actually do the polluting, but they are ultimately responsible.

This leaves industrial lobbies. There have been attempts in recent years to alert some of the winner sectors to their potential gains—IPPR has played an active role in this, and Friends of the Earth has also been building its links with business on the issue. It would be possible to buy off the opposition of the main losers—energy-intensive, highly traded sectors—by granting exemptions. But this would undermine the rationale for the reform, leaving the most polluting firms with little incentive to clean up, and would also reduce the revenue available to reduce distortionary taxation.

A better approach would be to allow a long lead time so that such sectors could adjust, and to design the package so that tax avoidance was possible for major energy users if they behaved in a sustainable manner. For example, it would be possible to exempt 'embedded' renewable generation from tax, making it worthwhile for, say, a large chemicals plant to develop its own source of electricity.

Finally, care should be taken to protect small and medium enterprises (SMEs). They will often be less well placed to respond to tax changes; most do not have dedicated environmental managers, for example. One way to help this sector would be to use some of the revenue to reduce business rates, an illogical tax which bears no relationship to profit or turnover, and which impacts heavily on small businesses.

A Green Tax Commission for the UK

Those countries which have introduced significant tax reforms along the lines proposed here have prepared the ground by establishing Green Tax Commissions, bringing together politicians (not just from the ruling parties), tax experts, industrialists, environmentalists and trade unionists. It would be sensible to do the same in the UK. This would ensure that the reforms were sensibly designed and implemented, and would help build support for the proposals.

Without a change of direction, environmental decline will manifest itself in public health effects that not even the most myopic of politicians can ignore. The social fabric cannot survive with existing levels of unemployment. Environmental tax reform would be a fitting project for any government, but should appeal particularly to a radical and progressive one.

Biographical Note

Stephen Tindale is Director of the Green Alliance. He was formerly Senior Research Fellow at the Institute for Public Policy Research and is the author, with Gerald Holtham, of Green Tax Reform: pollution payments and labour tax cuts (IPPR, 1996).

Environment, Risk and Democracy

ROBIN GROVE-WHITE

WHAT IS the relevance of environmental issues for wider political debates about institutional change in Britain? An understanding of the tensions surrounding currently dominant approaches to environmental problems is important not only for the sake of 'the environment' itself, but also in offering pointers to how the fabric and legitimacy of our political institutions more generally might be improved.

The recent history of UK environmental politics, and of certain specific *causes celebres*—the 1995 row about ocean dumping of the Brent Spar oil rig, the 1996 BSE crisis, and continuing grass-roots activism around new road-building in the UK—offer helpful prisms for understanding the limitations of current environmental policy approaches—and, more speculatively, for suggesting how the new challenges need now to be addressed.

What is meant by 'the Environment'?

First, a brief preamble. In most contexts, the term 'environment' refers to aspects of the material world. We acknowledge and discuss environmental issues overwhelmingly in *physical* terms—the state of wildlife species, habitats, landscapes, buildings, the atmosphere and oceans, and pollutants local and global.

But environmental issues are more than simply physical. They are also inescapably philosophical, ethical, political and cultural. The particular 'objective' environmental problems and issues which society recognises at any one moment are shaped and determined by processes of human judgement and social negotiation, *even in their very definitions*. In this particular sense, such issues are human 'inventions', their physical manifestations being mediated through human cultural 'filters' of many kinds.

The implications are far-reaching. One set of cultural filters may generate a quite different sense of what is at stake in the environmental 'field' from others. This means that the field vibrates with tensions arising from such differences of understanding. The contention of this chapter is that the version of environmental issues which is currently dominant in British political culture is still misleadingly narrow, and that widespread failure to recognise this is contributing to the nation's wider problems of political legitimacy and trust.

Popular environmentalism has developed in Britain since the early 1970s. As a phenomenon it has reflected mounting recognition of, and concern about, the accumulating physical 'externalities' of advanced industrial societies. But that is far from having been the whole story. Quite as significantly, if less obviously, environmentalism has also reflected a groping and often barely articulate reaction by many people against embedded patterns of thought and value (indeed, in philosophical terms, a dominant *ontology*) which appear to have been underpinning such problems. These dynamics can be seen in some of the issues which have emerged as most emblematic of environmental concern, as advanced by NGOs like Friends of the Earth, Greenpeace, the Council for the Protection of Rural England, the World Wide Fund for Nature and others over the past 20 years.

Take the issue of civil nuclear power in the 1970s and 1980s. The physical impacts of particular nuclear power-related developments were, recurrently, the apparent focal points of environmental controversy—levels of radioactive pollution from particular new plants, the risks of accidents, arrangements for nuclear waste transport and disposal, local landscape and community impacts, and the like. But wider underlying concerns were still more crucial. Where was the growing dependence on centralised technologies like nuclear energy leading society? On what basis could our political institutions be trusted to anticipate and provide against the *cumulative* unknown environmental and social feedbacks resulting from more and more plutonium, low level radiation and toxic waste generation? Need society pursue such modes of energy conversion, with their open-ended social and environmental implications, when potentially 'softer' technologies and social commitments could meet human needs quite as adequately? Such broader issues bubbled beneath the surface in a succession of controversies surrounding Windscale (now Sellafield), Sizewell B, and other nuclear developments from the early 1970s onwards. They reflected a sharply different perspective to that dominant in Whitehall and Westminster over most of the period. In this way, individual disputes about particular proposed nuclear energy developments came to possess far wider symbolic significance, as signals of mounting 'philosophical' concern about trends in industrial society.

Similarly with intensive agriculture. Behind recurrent specific brouhahas about threats to particular habitats, wildlife species, trees and hedges, or rows about the impact of chemical pesticides and fertilisers, has lain more profound unease at the cumulative implications of ever more industrialised approaches to food production, and at the increasingly manipulative and instrumental attitude towards nature they appear to imply.

Comparable patterns can be seen in numerous other spheres—transport issues, global climate change and depletion of the ozone layer, industrial waste disposal and packaging, urbanisation of the countryside. Issues have emerged into the public domain as particular physical problems. But frequently, behind this, they have been expressive of deeper cultural anxieties, reflecting inchoate unease at the hubris of industrial society's technological roller-coaster, doubts about the adequacy of our present patterns of knowledge and regulation, and concern at industrialism's apparent insensitivity towards many deeper human values and concerns.

The UK Government's Role

Government's role in these tensions of environmental understanding in the UK has been a central one. The Department of the Environment (DOE) was created in 1970, in the wake of the 1967 Torrey Canyon oil pollution disaster. From the outset, its approach to environmental problems sought to be managerial and apolitical—parsing and assessing particular problems so far as possible within frameworks of technical evaluation and formalised 'risk assessment'. In important respects, this reflected the belief within government that environmental problems were technical side issues, compared with the more mainstream political and economic priorities of modern societies.

Hence the 'rational' approach to environmental problems was to analyse and address them so far as possible in reductionist scientific and economic terms. This approach was reinforced by pressures internal to government—not least DOE's need for robust justification of what were perceived as the 'costs' of environmental improvement, against the scepticism of other more powerful Whitehall baronies such as the Department of Trade and Industry, and Treasury; and the need to argue Britain's corner in European Community negotiations. Overall the tacit assumption was that, consistent with Whitehall's long experience of 'top-down' dirigisme towards the public, such restrictive definitions of environmental issues—reflecting a form of environmental 'positivism'—would deliver the political goods.[1]

As the 1980s progressed, however, such rigidities in the Department's outlook towards environmental issues put it in constant tension with the wider environmental *weltanschauung* which was crystallising, in the UK as elsewhere, due to a combination of NGO efforts, international influences, and events. Repeatedly, on issues as different as acid rain, ocean dumping, green belts, whales and seals, Whitehall's supposed 'rational' approaches were experienced as grudging and untrustworthy, notwithstanding the conscientious efforts of individual Ministers and officials. Environmental NGOs became increasingly adept at framing their wider concerns in terms that resonated with the public whilst also being consistent with the Department's reductionist idioms.

In September 1988, a landmark speech to the Royal Society by Prime Minister Margaret Thatcher responded to the escalating popular and expert concern by projecting environmental issues—in particular *global* environmental issues—as both a mainstream geopolitical priority for government and as a central Whitehall preoccupation. In consequence, both the performance and the relative Whitehall standing of the Department of the Environment (in its environmental policy role) began to improve significantly in the early 1990s. Mounting international obligations flowing from the various 'sustainable development' conventions agreed at the 1992 UN Rio Earth Summit and from continuing European Union commitments helped consolidate such shifts. Nevertheless, despite the growing range and heightened significance of its activities, the Department's conception of the nature of environmental issues continued to be predominantly that established in the pre-1988 era—as a class of problems to be specified in physically reductionist terms, and then addressed through more or less existing power structures and relationships. Indeed, notwithstanding a succession of admirable post-Rio initiatives reflecting the Government's new commitment to 'sustainable development' (of which more below), engagement with the deeper cultural dynamics of environmental issues continued to be subsumed politically beneath Whitehall's characterisation of the problems as essentially physical, and thus, domestically at least, as tractable in principle to scientific, managerial and economistic methods of control.

A range of influences external to government helped consolidate these tendencies. Most media representations—TV and press particularly—tended to focus on *visible* phenomena and 'events', rather than on serious interpretation of the deeper cultural and political undercurrents of which these were manifestations. And—whatever their broader philosophical commitments— the major NGOs were drawn ineluctably to reinforce such tendencies towards media superficiality, in pursuit of continuing political impact. In consequence, with qualified exceptions (interestingly, the Conservative Secretary of State, John Gummer amongst them), politicians and the political parties tended not to challenge the entrenched positivism.

Thus overall, even in an era when UK government commitment to environmental values has never been higher, the gravitational pull towards a still-circumscribed representation of what is at stake has been evident. The particular intellectual tools relied upon by the Department of the Environment for its public authority—the doctrine of 'sound science', formalised risk assessment, and economic evaluation—reflect this reality.

'Audit' Dynamics

Before examining the ways in which recent environmental crises are now suggesting the alarming limitations of such approaches, it is important to highlight two recent background political developments which have been further reinforcing these tendencies.

The 1990s have seen an acceleration of major 'invisible' alterations in relationships between government and governed in Britain. Of particular note have been continuing changes in the organisation of the civil service, and intensified attempts to control public expenditure.

As regards the first of these, there has been a rapid spread of new forms of management in the public sector, modelled upon an image of methods in the private sector—the 'hollowed-out state' documented by political scientists such as Rhodes.[2] Around such changes, a previously unfamiliar Whitehall 'audit culture' (in Michael Power's term[3]) has been crystallising, involving ever-more attention to the 'delivery' of individually specified 'objectives', and the achievement of largely numerical 'performance targets', by a host of devolved agencies of government, amongst them many now responsible for activities undertaken previously on a more discretionary basis within the civil service. Whatever any wider benefits in terms of enhanced transparency and 'efficiency', such changes have been generating distinctive adaptive behaviours within the agencies themselves. The author's own experience of one such agency (as a statutory Forestry Commissioner) confirms the impacts of the new requirements on the *kinds* of priorities that officials within such agencies now tend to consider realistic (i.e. as potentially defensible in the face of Treasury and National Audit office scrutiny). There is increasing pressure to keep the targets objectively specifiable and discrete one from another, while also attaching numerical costs to them to the maximum extent possible.

It is not hard to see how, in the environmental protection sphere, such a new official monitoring philosophy (whatever its benefits) tends to reinforce the positivist approach to environmental issues described above. The new requirements push towards the characterisation of environmental goods and bads in atomised and reductionist terms, in order to render the associated expenditures amenable to 'independent' scrutiny and action.

Complementary to such new 'audit' dynamics—indeed, arguably their key drivers—have been progressively more intensive efforts by central government to rein in levels of overall public expenditure. These are reflected in recent changes to the annual PES negotiating processes, which have increased Treasury control and attenuated the bargaining power of non-Treasury departments. Again, the effect is to ratchet up continuously the pressure on the latter to specify in disaggregated, preferably monetary, terms policy commitments previously regarded as having lain outside such 'audit' vocabularies, other than in the most general sense (nature conservation obligations, for example). Such tighter Treasury and National Audit Office control is rippling through the environmental policy world. It helps explain, for example, the continuing enthusiasm within the Department of the Environment and its agencies for the development and use of economic tools ('contingent valuation' methodologies, for example), purporting to translate what is valued environmentally into a range of fragmented individual goods with numerical *money* values, despite mounting intellectual criticism of such methods. It also helps explain recent drives towards cost-benefit 'rationalisation' of public expenditures on risk management in the industrial and health spheres.

The cumulative effect of these various developments within government in the environmental domain has been subtle and pervasive. New and powerful administrative imperatives have been reinforcing the already limited ways in which the environmental field is characterised and addressed within government. But beyond this, as the priorities of 'audit' grow in significance for individual departments and agencies, there are signs of a shrinkage in the latters' capacity to listen and respond to new configurations of public concern and unease unrecognised by such idioms. Indeed, recent research is beginning to point to a relationship between such tendencies and the widely observed deterioration of public trust in government in Britain, a matter to which we return below.

Policy and Controversy

So far, this chapter has argued that a range of contingencies—historical, political and administrative—have been combining to produce a particular conception within government in Britain of the overall field of environmental issues (that is, of what sociologists refer to as the *problematic* of the environment). Despite the high degree of commitment amongst key officials and Ministers, this conception continues to be restrictive and positivist, reinforcing a continuing underlying emphasis on the physical manifestations of particular problems at the expense of genuinely creative political and intellectual engagement with the mounting cultural challenges to which these problems are pointers.

The resulting tension in Britain between the reality (as the writer would see it) of 'the environment' as a key arena of struggle around the central trajectories of advanced industrial societies and their patterns of technocratic governance, and the narrow positivist understandings so obstinately internalised within government, are emerging in the form of increasingly obvious problems in the public policy domain.

Indeed, as the stakes for government have become higher, and mainstream political commitments to addressing environmental problems have become increasingly ambitious—witness the mounting range of White Papers, policy initiatives and international agreements to which the UK Government is now committed—so the limitations of the official characterisation of the field are becoming increasingly clear. Not only is the orthodoxy beginning to generate more rather than less political controversy and increasing costs in key domains, it is also contributing, apparently innocently, to an escalating corrosion of public trust in official institutions more generally.

The Brent Spar

Three recent controversies point to the increasing difficulties the polity now faces in this regard. In each of the three, the fragility of a particular key bulwark of the current official approach can be seen.

The first concerns the episode of the 'Brent Spar' and its implications for government's reliance on the axiom of 'sound science'. In this case, Shell's proposal to sink a massive redundant oil facility in the Atlantic approaches was frustrated by a swingeing consumer boycott, triggered by Greenpeace's spectacular occupation of the rig in July 1995.

Through the media, Greenpeace converted the Brent Spar platform into, effectively, an environmental icon—a compelling symbol for the cavalier use of the oceans as an industrial dump should such disposals be allowed. Shell

and the Government argued that all the appropriate environmental evalua-
tions, backed by 'sound science' under the official 'Best Practicable Environmental Option' procedures, had cleared the proposal. Nevertheless, after
Shell had been forced to back down, a measured analysis by the Natural
Environment Research Council concluded that both Shell *and* Greenpeace had
each had a sound case in their respective terms.

A crucial feature of the episode was its stark demonstration of the fragility
of the then Government's 'sound science' justification for the proposal. The
Government's cultural filters (to use a metaphor from the introduction to this
chapter) had stipulated that such disposals should be evaluated scientifically
on a *case-by-case* basis. By contrast, Greenpeace's filters implied the need for a
much longer view, reflecting the *cumulative* implications of the series of
subsequent dumpings that would follow this first authorisation. Later
surveys suggested that a majority of the public shared Greenpeace's perspective, and regarded the narrower official science as untrustworthy.

For government bodies committed to a positivist perspective on environmental issues, the ability to stipulate what does or does not constitute 'sound
science' in a particular case had previously been a crucial and resilient source
of political authority. This being so, the significance of the Brent Spar case lay
in its demonstration of the chronic intellectual vulnerability of this assumption, when subjected to politically imaginative challenge. The proper lesson
was not that public 'irrationality' prevailed over official objectivity (as
Ministers claimed angrily after the event), but rather that in the real world
all 'scientific facts' are, necessarily, the products of prior *social* commitments,
which predetermine the appropriate boundaries of what is held to be at issue.
Government's insistence on the unique 'rationality' of its own restrictive
framing of the issue, so at odds with wider cultural perceptions (and with
contemporary scholarly understanding of the nature of science itself), simply
intensified public mistrust of its good faith.

However, not everyone missed the point. A striking feature of the post-
Brent Spar fall-out has been the pragmatic seriousness with which senior
figures in industry (in contrast to those within government) have responded
to the outcome. There appears to be mounting recognition in a number of
large companies of the need no longer to assume that the government's
appraisal methodologies are robustly grounded, and to begin instead to
engage more directly with broader *social* currents of concern.

BSE

Parallel issues arise from a second illustrative case—that of the continuing
crisis of BSE ('mad cow disease') and its implications for human health. If
Brent Spar can be seen as having been a watershed for the authority of official
doctrines of environmental 'sound science', BSE raises equally serious
questions about dominant official approaches to the identification, assessment
and management of environmental *risks*.

Again, the authority of the established approach to environmental policy in Britain rests on the claim to be able to anticipate and guard against new forms of environmental hazard. The methodologies relied upon by the Department of the Environment, Health and Safety Executive and various ministerial advisory bodies for this purpose are predicated on the assumption that they can establish and quantify the broad probabilities of occurrence of particular hazards. Such methods work well within a substantial range of closed systems (many industrial plants for example) where pathways and processes are largely known and understood.

However, in the BSE case, the system in question is open-ended and full of unknowns. Human intervention in nature (in this case, involving a radical change to the food patterns of herbivores) has produced inadvertently a potential risk to human health, the seriousness and boundaries of which remain intrinsically impossible to quantify. In this particular sense, BSE can be seen to stand for any one of a mounting range of possible hazards now being generated as unintended consequences of dynamic processes of technological innovation in industrial societies like our own. (Numerous other topical examples might be cited—for example, the disturbing unknowns surrounding omnipresent hormone-disrupting chemicals and new pesticide-resistant bacteria).

As the BSE saga has shown, the techniques of formalised risk assessment relied upon by government offer little proactive intellectual purchase on such open-ended phenomena. Indeed, the spectacle of media spokesmen for ministerial advisory committees trading on the inherited authority of scientific objectivity to provide casuistical public reassurance during the political turmoils triggered by the crisis has been corrosive of public confidence in both science *and* government.

Yet the positivist framing of environmental issues to which British institutions of government appear to be committed means that reliance on such assessment processes continues to be insisted upon as the 'rational' approach—with the corollary that in practice firm indications of harm are deemed necessary *before* preventative action can be justified (whatever government's repeated endorsements of the so-called precautionary principle).

An emerging body of contemporary social analysis around the issue of 'risk society', by Beck, Wynne, Giddens and others,[4] suggests that the environmental problems of the future will increasingly involve such unpredictable and intrinsically incalculable risks, placing huge burdens of public trust on society's processes of risk regulation and negotiation. Seen in this light, the BSE crisis, with its immense political and economic costs (£3bn and rising), is a deeply worrying portent for the future. It highlights again the disturbing inadequacy of the positivist model of how such issues are to be understood.

Road-building

A third and quite different example of the limitations of this officially-embedded model relates to transport policy and the countryside. Here too the official doctrine has been that significant environmental issues are those which can be pictured in objective and measurable terms, as externalities arising from more mainstream commercial and political necessities. Over the past 25 years this doctrine has been embodied in the economistic techniques of Cost Benefit Analysis. This has been employed by the Department of Transport effectively to marginalise deeper environmental concerns. Cost benefit analysis has been used to legitimize massive road-building, escalating national dependence on motorised mobility and relative neglect of public transport, the latter probably compounded by recent privatisations of bus and train systems.

In the 1970s and 1980s, NGOs like Friends of the Earth offered powerful intellectual critiques of most of these tendencies—critiques which more recently have been echoed and further elaborated by the Royal Commission on Environmental Pollution. Little has changed, however. Official road traffic forecasts point inexorably upwards, complementing already emerging signs of gridlock in the South East of the country.

However, back to back with this burgeoning crisis, new patterns of dissent and symbolic direct action have been emerging, to growing public approval. The tree-dwellers and tunnellers of the Newbury by-pass, A30, and other campaigns—scorned and patronised in Whitehall and Westminster, but nevertheless reflecting many people's frustration at the human and cultural insensitivity of recent policy directions—are graphic new bearers of the message that the officially embedded positivist model is chronically unable to do justice to the deeper cultural issues at stake for society in the field of 'the environment'.

'Sustainable Development'—and Beyond

What then is the way forward? In crisis lies opportunity. It is important first to accept that on present trajectories, unavoidable and increasingly difficult environmental policy decisions lie ahead. As a recent Secretary of State acknowledged, we are entering an era in which the easy environmental plums have been picked, and in which little future progress will be made without painful costs to many established interests in society. All of us now stand to be affected. For example, on present trends, reducing fossil fuel pollutants from industry and transport in order to help mitigate the mounting prospect of global climate change, will almost certainly require unpopular new forms of regulation or taxation. The bitter political rows surrounding the introduction of VAT on domestic fuels in 1994 give a modest foretaste of what may be in store. Similarly, civilising the motor car and reinvigorating public transport will bring substantial costs and inconveniences to most of us, in order to avoid longer-term chaos for all. And so on. Policies of such kinds will only be achievable if deep and resilient public support can first be generated, rooted in a shared conviction that such commitments are the just and creative way forward for the population as a whole.

But as this chapter has argued, the political conditions for such consensus appear dangerously absent at present. Not only has the dominant discourse on environmental issues evolved as a technocratic and publicly unappealing one which, through its increasingly conspicuous limitations, now risks alienating people from socially desirable action. But quite as serious, the surrounding social *zeitgeist* is one in which appeals by political leaders for concerted public action now appear to be regarded with deep cynicism and mistrust. Indeed, it is a climate in which there are deep and widely shared doubts about the adequacy and legitimacy of our established political institutions and mechanisms. So what needs to be done?

The first priority is to advance and crystallise a far wider recognition of the central significance of the environmental 'field' as the key prism through which the deeper cultural tensions and socially creative opportunities in contemporary society are beginning to be addressed. This is hardly a new or original suggestion, but it needs to be urged with fresh energy and imagination. The continuing strength and vitality of environmentalism, in myriad different registers and associational forms, is a truly remarkable feature of contemporary Britain, particularly when set in the context of the benignly dead hand of the dominant positivism. But while such flourishing testifies to the resonance for many of environmentalism's intuitions about the present and future world, it has to be recognised that the large majority of people connect with it in their day-to-day lives barely at all. Indeed, recent qualitative research studies in various parts of Britain have found that whilst many people attach great value to their local environments, and are deeply concerned (but fatalistic) about global environmental issues, they are alienated from and scornful about the discourses and activities surrounding *official* promotion of environmental values.[5] This is sobering and points to the distance that has yet to be travelled.

Imaginative steps have already begun to be taken. The initiatives on which the Government has set most store are those arising out of the 1992 Rio Earth Summit's accords on 'sustainable development'. These hold some promise, and are generating a range of creative responses inside and outside government. But their current limitations need to be understood as much as their strengths.

Three key initiatives launched by the then Prime Minister in early 1993 following Rio are of particular significance. The first is the 'British Government Panel on Sustainable Development', a five person committee of serious intellectual quality and judgement, chaired by Sir Crispin Tickell, formerly UK Ambassador to the UN. It reports directly to the Prime Minister, and is now emerging as a significant influence in Whitehall, calling publicly to account all arms of government in furtherance of the latter's sustainable development commitments. Reports suggest that it is beginning to become an effective catalytic presence within official circles, albeit in the traditional insider mode of British governance. The second is the Sustainable Development Round Table, chaired by the Secretary of State, and involving significant figures from government departments, industrial companies, universities and NGOs in discussion and negotiation about policy advances. It has a civil service secretariat, and has begun to produce useful critiques of policy sectors where action to further sustainable development needs more attention. The third national initiative is Going for Green, which embodies both a national publicity campaign and a series of experimental pilot community initiatives aimed at fostering grass roots commitment to sustainable development objectives such as reduced household energy consumption and greater use of public transport. It seeks to complement the proliferating range of modest sustainable development initiatives at local authority level promoted under Local Agenda 21, a further post-Rio initiative.

All of these are significant innovations. They are engaging the energies of a range of already committed individuals and institutions in pioneering ways. But so far there is little sign of a breakthrough into the wider public consciousness.

The logic of the arguments developed in this chapter is that this difficulty reflects the extent to which all of the various initiatives, however visionary in their 'sustainable development' aspirations, continue to be hobbled by the limited positivist framing of the environmental problematic in which they are embedded. This is least surprising in the cases of the Panel and Round Table—since both are designed to address public authorities and industries which broadly endorse such framings. But a different story is beginning to emerge from on-the-ground experiments under Going for Green and Local Agenda 21—in which new forms of direct engagement and interaction with the wider public are being attempted. Academic social researchers associated with Going for Green's six DOE-funded 'Pilot Communities' experiments around the country (of whom the present writer is one) are being brought face to face with the striking lack of resonance of the dominant model of environmentalism for ordinary people's lived experience. This confirms other recent research, and is a finding of considerable national significance. For it implies, disturbingly, that present official approaches to harnessing the wider public's interest in sustainable development may simply be inadequate for the job. It points instead to the critical and urgent need for *fresh* ways of connecting with latent public energies, if serious coalitions for change and collective responsibility are to be generated.

Civil Society and Government

The way forward lies within civil society itself. Britain currently has an immense wealth of dynamic and creative activity by its citizens, outside formal institutions. The confidence and imagination propelling new networks and informal patterns of cooperative association in many sectors and levels of British society is a well of huge social resilience and inventiveness. Our current creativity in the worlds of new theatre, film and music is paralleled by the explosive proliferation of ever more energetic self-help networks in the fields of illness and social action, of new cultural movements around phenomena as diverse as leisure, animals, sport, crafts, 'alternative' medicine, therapy and religion, ethnic identity, regional identity, youth culture, and simply lifestyle, not to mention the wealth of NGOs in practically every domain of public life. As Bronislaw Szerszinski argues in this volume, such rich and varied patterns of associational life in Britain—unique, perhaps in Europe—are a phenomenon of immense potential significance for the future of the polity generally, and for progress towards 'sustainable development' in particular.

Moreover, a similar depth of citizen imagination is now manifesting itself as part of the reaction to recent convulsive events in the sphere of environmental risk, such as the Brent Spar and BSE crises. The growing range of interactions, and even of prospective coalitions, between particular industrial actors and NGOs across previously unbridgeable divides, boosted by reflection on the Brent Spar episode, appears itself to hint at modest but potentially significant reconfigurations within civil society. So too does the intensity of emerging debate about the nature and role of science in society within independent 'establishment' institutions such as the Royal Society, Royal Society of Medicine, the Research Councils, the Fellowship of Engineering and the like, in the wake of recent risk crises such as BSE. Though the politically hard-bitten may view such developments with scepticism, I suggest it may be wiser to understand them in more positive terms—as early stages in a new and painful opening of minds, in domains where long established political and intellectual hegemonies have previously reigned unchallenged. Crucially, these are changes which have been triggered by processes of *democracy*, in which citizen (or 'consumer') power has begun to catalyse structural shifts in formerly stable social arrangements, independently of conventional political processes. If this is right, it heralds a development of genuine significance for the country's future.

The precise institutional implications of these phenomena for government, and more immediately for environmental policy and sustainable development, need now to be clarified. The burgeoning range of associations in civil society *outside* the overtly environmental domain contains many clues as to how the creative energies and wider sensibilities of an increasingly diverse public are now developing. It is with these that environmental policy and steps towards sustainable development must now engage and build, if authentic public identification is to be advanced.

It follows that the environmental policy community inside and outside Whitehall and Westminster must learn to *listen*. The 'hollowing-out' of the state has many merits, but only if the interventions which the state *does* initiate are then attuned with genuine sensitivity to what is going on in the sinews of society itself. The argument of this chapter has been that, in the environmental sphere, so crucial to our futures, this attunement is proving damagingly inadequate, and that recent crises—Brent Spar, BSE, motorways—have begun to signal this.

In the immediate future, the need for such new *listening* may point less to a requirement for major institutional upheaval—the Charter 88 model, so to speak—than to a need for more and better *social intelligence* in government about the moral and cultural mutations now occurring outside established policy communities, and about the potential synergies these may offer for progress in advancing sustainable development aspirations.

This is largely new terrain for Whitehall, and indeed for environmental 'policy community' actors more generally. For it points to an urgent need for frameworks of understanding lying outside the particular 'positivist' idioms chided in this paper—towards a need to build, perhaps, on qualitative and relational insights of the kind now emerging from contemporary anthropology, action arts, and even religious praxis. The new international idioms of independent citizen juries, discussion panels, town meetings, consensus conferences and the like, now being highlighted by IPPR, the Centre for Policy Studies, the Science Museum and others,[6] offer promising arenas for experiment. Bigger, more structural reforms need to be designed and calibrated to reflect the refreshed understandings which could be encouraged to emerge from such processes. Meanwhile, central government should play an energetic but tactful hands-off *enabling* role, in ways that encourage and do not impose upon the energies now being released.

For reasons peculiar to Britain's recent history and winner-takes-all electoral system, well-articulated by David Marquand in his 1988 study, *The Unprincipled Society*, issues of these kinds have attracted surprisingly little mainstream attention from the two major political parties. That too will have to change. The clues for greater optimism, creativity and national 'success' are there to be discovered in the environmental field and in civil society more generally.

There is little time to be lost.

Biographical Note

Robin Grove-White is Director of the Centre for the Study of Environmental Change, Lancaster University. He is also currently a Forestry Commissioner, Chair of Greenpeace UK, and a board member of the Green Alliance and Common Ground.

Notes

1 R. Grove-White, 'Environmentalism: A New Moral Discourse for Technological Society?', in K. Milton, ed., *Environmentalism: The View From Anthropology*, London, Routledge, 1993.
2 R. Rhodes, 'The Hollowing out of the State: the Changing Nature of Public Service

in Britain', *The Political Quarterly*, Vol. 65, No. 2. Also C. Foster and F. Plowden, *The State Under Stress*, Buckingham, Open Univeresity Press, 1996.

3 M. Power, *The Audit Explosion*, London, Demos, 1995.

4 U. Beck, *Risk Society: Towards a New Modernity*, London, Sage, 1992; A. Giddens, *The Consequences of Modernity*, Cambridge, Polity, 1990; B. Wynne, 'Scientific Knowledge and the Global Environment' in M. Redclift and T. Benton, eds., *Social Theory and the Global Environment*, London, Routledge, 1994.

5 P. Macnaghten, R. Grove-White, M. Jacobs and B. Wynne, *Public Perceptions and Sustainability in Lancashire*, Lancashire County Council, Preston, 1995.

6 J. Stewart, *Further Innovation in Democratic Practice*, School of Public Policy, University of Burmingham, 1996.

Using Science

PHYLLIS STARKEY

Scientists and Society

My contribution to this book is an attempt to explore the interface between science and public policy in respect of the environment, from the point of view of the scientist.

The first, fairly obvious, point to make is that scientists are not a homogeneous group; like any other professional group their behaviour and their opinions are shaped by their experience, both as scientists and as people. The only thing that unites them is their initial training in a scientific methodology which implies a certain analytical and critical way of thinking, and which relies on measurable or quantifiable evidence against which hypotheses are tested. Scientists divide themselves by subject, by whether they are involved in theoretical 'pure' science or more applied science, and by the sector in which they work. Scientists working within industry, in research institutes, in universities or involved in policy making within government departments or other public bodies, all have different viewpoints. Finally, scientists are also people and can be divided, as can most other groups, into the activists, a relatively small number of people using their scientific knowledge to participate in wider public debate and action, and the vast majority who just get on with their own lives, oblivious to the public debate until it intrudes on their own personal interests.

Within this heterogenous population of scientists, the group I wish to focus on are those involved in research in the public sector, particularly in universities. This is partly because they are the group of which I have most experience and partly because it is this group whose expertise is most often cited in public debate. It is also this group that has been subject recently to arguably the greatest changes in their working environment.

John Ziman[1] has described the factors he believes are reshaping the practice of scientific research within universities, leading to what he calls a 'post-academic' science. This is characterised by a more collective and less individualistic approach, where knowledge is more directed towards solving specific problems rather than being regarded as an end in itself. The greater complexity and sophistication of scientific research means that scientists are increasingly working in larger multidisciplinary groups, focused on common specific aims. Even in the biological sciences, research is increasingly characterised by 'big science', as the sophisticated equipment and range of research skills required to do leading-edge research demand long-term and large-scale funding. Similar pressures are operating on industrial research

© The Political Quarterly Publishing Co. Ltd. 1997
Published by Blackwell Publishers, 108 Cowley Road, Oxford OX4 1JF, UK and 350 Main Street, Malden, MA 02148, USA   123

with technological developments also generating heterogeneous hybrid teams and encouraging increased collaboration between academic and industrial scientists.

The changing nature of the scientific enterprise is reflected in the government White Paper on science, engineering and technology 'Realising Our Potential.'[2] Research councils, the main source of public sector funding for university science research, were given a clear mission to support research and training in order to contribute to the UK public good through increasing wealth creation and improving the quality of life. In particular, public funding of basic research is justified in terms of the need to maintain a seed-bed from which can arise the unforeseen technological advances that industry can exploit. Research council priorities were to be developed to meet the needs of users and beneficiaries; for the BBSRC (Biotechnology and Biological Sciences Research Council), for example, those users and beneficiaries were spelt out as including the agriculture, bioprocessing, chemical, food, healthcare, pharmaceutical and other biotechnological related industries. At the same time, the underpinning funding of university research infrastructure through the Higher Education Funding Council, was subjected to greater competition and selectivity. Alternative sources of funding from medical charities and industry are similarly focused on basic or applied science related to their particular disease or industrial interest.

The increasing emphasis on science for a purpose has raised new questions. For the research councils it has posed the problem of how, as public bodies assessing the needs of users and beneficiaries of their science, they should take account of the public as user and beneficiary. Most research councils have found it relatively straightforward to set up mechanisms for consulting with industrial users, and integrating their views into policy making, even if real involvement of a wide range of industrial users is more difficult to achieve. Much more problematic is how a research council can take account of the public, the citizens' views, in assessing scientific priorities, in deciding what research would best contribute to wealth creation and the quality of life. Are government users, and for BBSRC this would be primarily MAFF and the Departments of Health and the Environment, the route for expressing public demands, or should there be a more direct way of integrating public concerns into research council policy?

The Experience of BBSRC

The science within BBSRC's remit covers a range of basic, strategic and applied research relating to the understanding and exploitation of biological systems. It is an area of science that is having an increasingly important role in stimulating significant change in industry and in our way of living. Many of the options that arise from research in the biological sciences, and in biotechnology in particular, arouse considerable public debate. There has been a shift of public concern from physics to biology. Nuclear power, whether for military or civil use was the great battleground of the 1950s and 1960s. Today's issues are much more concerned with genetic modification of plants and animals and manipulations of human genetics and human reproductive technologies. Once science starts to affect the food we eat, the animals and plants around us and our own bodies public interest and concern is bound to be heightened.

For BBSRC, the public concern has focused on the environmental and ethical implications of the genetic modification of plants and animals, the welfare of both experimental laboratory and farm animals, the desirability of research to underpin intensive agriculture, animal husbandry practices or crop protection regimes that may have unforeseen effects on human health and the environment; food safety and quality related to the needs of consumers as opposed to those of food manufacturers; and substitution of food crops by genetically modified plant crops to provide renewable sources of fossil fuel substitutes.

The Public's Understanding of Science

The intensity of public debate has demonstrated that scientific priority setting is a matter of public concern. There is a perception among scientists that science and scientific advances are increasingly seen by the public as a threat rather than as progress contributing to a better quality of life. Scientists are increasingly on the defensive. This has led to the view within the scientific community that 'something needs to be done' to improve the public's understanding of science. Indeed 'providing advice, disseminating knowledge, and promoting public understanding' has been incorporated into the mission of each of the research councils. Informing the public, in non-technical accessible language, about science and its potential to change the way we live is important if the public is to be able to participate in any sort of debate. Programmes of Public Understanding of Science aimed at school children are particularly helpful in allowing scientists to communicate some of the real excitement that attracts most scientists to science in the first place. But for too many scientists, there is a tendency to think that if only the public understood more about science then controversy would cease. Lewis Wolpert's view[3] that 'science is central to our culture and difficult for non-scientists' preserves the superiority of scientists as experts and by definition particularly clever people who have been able to understand science, but suggests that we simply need to persist in educating the public, rather like shouting at foreigners, until they see the point.

A recent article by Dr Richard Sharpe[4] whose research results were the trigger for the 'phthalates in baby milk' story is very revealing. Dr Sharpe described how in concluding that the risk to the public from phthalates in baby milk was minimal, he took 'into account many other factors which I as an expert know about'. In his view the public reaction was driven by journalistic inaccuracy and the inability of laypeople to judge relative risk.

An alternative analysis might indicate that the public too have some expertise and that their assessment of risk was influenced by the fact that precise information on the levels of phthalates in different brands was being withheld and that past experience, from radioactive contamination around air bases to BSE, suggested that government only withheld information when there were strong grounds for public concern. A greater awareness by scientists of the work of social scientists on the factors affecting public perception of risk might enable scientific experts more accurately to assess the likely public response to their pronouncements and modify their approach accordingly. It may also help scientists to see why what they say is perceived differently depending on whether it is cited in support of a government, industrial or pressure group viewpoint. The public's level of trust of an institution will colour their perception of the disinterestedness and reliability of any scientific advice associated with that institution.

Scientists Understanding of the Public

There is, however, beginning to be a recognition in the scientific world that in the dialogue between the public and scientists the communication problems are not all one way; that scientists need to understand their limitations as well as their strengths; that science is not value-free but operates within a cultural and social context. It is accepted that scientists need to be clear about the areas where they do have expertise and those where they do not, and that 'scientific expertise is contestable'.[5]

This recognition has led to developments in promoting more effective debate, and many of these developments borrow experience from other countries. BBSRC for example attempted to stimulate a public debate on the ethical issues relating to the application of genetic modification of plants through a Consensus Conference on plant biotechnology in December 1994. It was organised by the Science Museum to ensure its independence, and was modelled on the consensus conferences pioneered in Denmark and The Netherlands. The aim is to allow lay people to engage in a thorough discussion with experts, to make recommendations and thus to reveal some of the prevailing ideas and concerns of citizens. To achieve this the lay panel needs to be as representative as possible.

For the UK Consensus Conference the panel of 16 lay people was chosen from those who responded to a public advertisement. Care was taken to select a wide cross-section, both women and men, with a spread of ages, educational level and occupation and from all parts of the country. The panel members took part in two briefing weekends, then three days of interviewing experts, including environmental activists, before setting out their conclusions. The panel members framed the seven key questions around which the conference was organised, identified the key expert witnesses they wished to call, and wrote the final report which was issued in preliminary form on the final day of the conference.

Members of the lay panel found the whole experience to be very positive. Faced with often conflicting evidence from the experts, particularly on the question of the possible impacts of plant biotechnology on the environment, their recommendations were largely supportive of the development of plant biotechnology but stressed the importance of retaining public trust through openness, a proper caution on the part of scientists and industry, and rigorous UK and European regulation. In the light of recent controversy about the inability to identify foods containing genetically-modified soya, it is significant that the lay panel strongly recommended clear and meaningful labelling of all genetically-modified foods so that the public could 'freely exercise its right to choose'.[6] From the point of view of BBSRC, the Consensus Conference, which was part of the Council's programme on the Public Understanding of Science, was successful in provoking considerable media interest and stimulating wider public debate at least for a short time.

The Consensus Conference explored the potential uses of plant biotechnology as well as any intrinsic concerns in the technology itself. A difficult area for many scientists is how far basic scientists need to concern themselves with the implications of the eventual use to which their research may be put. Many scientific developments can be used both for good and for evil and not unnaturally most scientists focus on the good. Plant biotechnologists, for example, are more likely to stress the contribution new drought-resistant crop varieties can make to reducing hunger in developing countries than pointing out how crops modified for growth in more temperate climates may, by substituting for tropical cash crops, actually increase Third World poverty.

A system of Technology Assessment developed by BBSRC's equivalent in The Netherlands, the DLO, helps scientists at least to consider the potential consequences of their research at the time when they are developing new research proposals. The DLO funds a number of research institutes involved in research underpinning agriculture, forestry and fisheries. Their Technology Assessment system encourages institute scientists to consider the social, economic and moral consequences of their science during the process of formulating new research projects. DLO scientists are asked to consider the likely affect of their proposed project on a variety of potential target groups including producers, suppliers, trade unions, environmental organisations, animal protection groups, consumers, religious organisations, central and local government, political groups, Third World and international organisations. A structured questionnaire guides the scientist through a process of assessing the benefits and disbenefits of the proposed research to each group. Researchers are encouraged to seek guidance where required by consulting social scientists, economists and others. The process is intended to encourage the scientists to design research projects which meet the wider needs of society and which are likely to be ethically and socially acceptable rather than formulating projects in a vacuum and being surprised by unanticipated public antipathy. BBSRC will be piloting a simplified version of the DLO questionnaire for scientists applying for funding for research involving the genetic modification of animals. This approach encourages individual scientists to think through the likely alternative outcomes from their research, even where the decisions on those outcomes are controlled by others.

As well as improving dialogue between individual scientists and the public, BBSRC has been considering how to incorporate more effectively an appreciation of public views within its own priority setting process. At present, BBSRC consults with the research community to identify 'scientific opportunities', that is the areas within the biological sciences where new and exciting developments are occurring. This is matched by consultation with users, including industry, relevant government departments, and environmental and consumer groups to identify 'market needs'. By bringing together the scientific opportunities and the market needs, BBSRC identifies its priorities for scientific spending for the future. Recent controversies, of which BSE/CJD is the most obvious, have highlighted the absence from this process of any wider consideration of the social-cultural context within which BBSRC is operating. To try to plug this gap, BBSRC has set up a small group including social and political scientists, consumer experts, retailers and industrialists to identify trends in public concerns relevant to BBSRC science and to comment on the social and cultural context of the scientific priorities identified by BBSRC's science board. It is hoped that this will inform BBSRC scientific priority setting.

Some action is therefore being taken to encourage more effective dialogue between scientists and the public. But the extent to which the public should be involved in decision-making is much wider than science: it is an issue for society as a whole. The Plant Biotechnology Consensus Conference had only limited influence because in the UK there is no mechanism for integrating such deliberations into the normal process of public decision-making. In Denmark, consensus conferences are organised by an independent body closely linked to a Parliamentary committee. The consensus conferences take place in the Parliament building, and at the end of the conference the final report is handed to a delegated MP and all MPs receive a copy. The 1989 report on the use of human genetic information for personal employment and insurance purposes directly led to action by the Danish Parliament to prevent such uses. Consensus conferences in the form of citizens' juries have been used in local government in this country to inform local decision making; but there are no comparable mechanisms at a national level.

A further cautionary point. The unspoken assumption of this book appears to be that increased democratic participation in decision making is bound to lead to 'greener' policies. That may be an unwise assumption. The majority public view may well continue to be that the current level of road deaths is 'a price worth paying' for the individual mobility that is dependent on widespread car ownership. Keith Tester, Professor of Sociology at Portsmouth University,[7] has argued that in the conflict between environment, biotechnology and animal welfare, there is evidence that public concern about animal welfare and the environment is simply a surrogate for their concern about the quality of their own food and their own immediate quality of life. Thus much public opposition to nuclear power was a response to concern about levels of radioactive isotopes in children's milk. Where biotechnology is perceived to improve human health, for example in the production of pharmaceutical proteins in sheep's milk, environmental and animal welfare concerns may well be downgraded. As scientists have found out, the public does not necessarily agree with you just because you have explained the 'facts' to them more clearly.

Biographical Note

Dr Phyllis Starkey is a biochemist. From 1970 to 1993, with two breaks when her daughters were young, she pursued a career in medical research. For eight years she was a University Lecturer in Obstetrics & Gynaecology and Research Fellow of Somerville College at the University of Oxford. At the time of writing she was Head of Assessment Branch for the Biotechnology and Biological Sciences Research Council. Dr Starkey is M.P. for Milton Keynes South West.

Notes

1 'Is science losing its objectivity?', *Nature*, 1996, 382, pp. 751–4.
2 *Cm* 2250, London, HMSO, 1993.
3 'Woolly thinking in the field', *The Times Higher Education Supplement*, 31 May 1996.
4 'How my work triggered the milk fiasco', *The Daily Telegraph*, 5 June 1996.
5 Harry Collins and Trevor Pinch, *The Golem: what Everyone Should Know About Science*, Cambridge, Cambridge University Press, 1993.
6 UK National Consensus Conference on Plant Biotechnology; final report, Science Museum, London, 1995.
7 Perspective, *The Times Higher Education Supplement*, 9 August 1996.

Making Environmental Policy

DEREK OSBORN

THIS chapter first outlines a classic rationalist approach to environmental policy-making based on sound science and good economics, and discusses the strengths and limitations of this approach together with some of the criticisms that have been made of it. It illustrates these points with some practical examples of policy-making and policy failures. Finally it raises some questions about the possibility and feasibility of working towards a broader synthesis as a basis for policy-making in the future.

The Rationalist Approach

The classic approach to environmental policy-making adopted by most Departments of the Environment in the world and underlying most international negotiations can be summarised in the familiar rubric of the 'state, pressure, response' model. This model provides a convenient summary of the three primary tasks of environment policy-making as being: to identify and monitor relevant states of the environment; to identify and analyse the pressures on it which may, if unchecked, lead to deterioration; and to devise appropriate measures of response to prevent or minimise adverse effects and, if possible, enhance the environment in the most economically efficient way.

Not only is this a well entrenched and rational approach. It is also one well suited to the capacities of the bureaucratic structures of Departments of Environment around the world, and the scientific and policy analysis communities which enjoy a symbiotic relationship with them. According to this model we cannot rely solely on the supposedly unscientific and amateurish perceptions of individual citizens and communities to understand the state of the environment and monitor it. It must be measured and assessed with scientific rigour and precision in research establishments and laboratories. Monitoring changes requires long-term time data series and full use of the most modern technologies in space, in the atmosphere, in the oceans and on the ground. Analysing the complex causes of changes in our environment is another major source of scientific enquiry, and vast co-operative scientific enterprises.

Analysis of policy options for responses is another fruitful area of development for professional expertise and the related policy communities, this time relying more on the efforts of economics, sociology and legal professions to supplement those of the scientist. The pure rationalist would like to reduce the whole decision-taking process to a complex cost-benefit analysis.

 Published by Blackwell Publishers, 108 Cowley Road, Oxford OX4 1JF, UK and 350 Main Street, Malden, MA 02148, USA

Strengths and Limitations of the Rationalist Approach

We should not reject all this in toto. The world *is* complex. The interaction of multiple causes producing environmental effects is very deep and sometimes counter intuitive, needing all the efforts of many different types of scientific enquiry to comprehend. The different options that may be available for dealing with an environmental problem or situation are very wide ranging and have different patterns of costs and benefits which need careful weighing, both in the aggregate and in their distributional and social effects. All of these matters must be thoroughly considered and assessed in establishing a just and effective environmental policy.

Nevertheless, there are some serious limitations to the whole approach which it is important for policy-makers and the whole policy community to bear in mind. First, we cannot know everything and certainly we cannot monitor everything. We have to make judgments about what to assess and what problems to explore. Science and scientific risk analysis may give us some guidance, but it can also be very misleading if it attempts to reduce everything to a quantified analysis, and pays too little regard to what real people and real communities feel and think are the salient features about their own environment. What people believe and what they care about need to be factored into the assessment of what is important.

Public Opinion

This is easier said than done of course. But there is nothing like trying. Those progressive local authorities all over the world that have pioneered new ways of engaging with their local communities to find out what people really wanted to be done to and for their environments have frequently surprised themselves with what they learned and the new ways of delivering services or planning programmes that emerged. Conversely, the folly of pressing ahead with controversial decisions on the basis of 'rational' analysis alone with inadequate consultation with the public and those affected is sadly demonstrated with monotonous regularity by public bodies of all kinds at all levels. The debacle of the recent Brent Spar episode is one example. When Shell with government backing tried to dump a disused oil-platform in the Atlantic, on the basis of a supposedly 'scientific' assessment that this was the best practicable environmental option, reckoned without the overwhelming intuitive opposition of the public. This was only the most spectacular of recent cases of the dangers of relying only on science and discounting public opinion.

The solution in that case, as in all the other embarrassing episodes that went before it, is so simple that it is astonishing that the lesson has still not been properly learned: 'Consult openly, widely, publicly and in advance with everybody who could conceivably be affected by a controversial decision. Go through public inquiries if necessary. Trust the people. Listen to what they say.' In other words, establish a proper process for handling such decisions.

In England the Department of the Environment has learned this lesson the hard way. Among government departments, it has the most well-developed processes and administrative instincts for handling issues and decisions in a comparatively open and consultative manner leading to decisions that stand up better to public scrutiny and challenge. It is noteworthy that several of the most recent bad experiences for Government, such as the Brent Spar case, or the BSE imbroglio, stem from administrative processes that were handled by other Departments in a much less open and consultative way, with too much weight being given to supposedly 'objective' scientific advice. This turned out in the end not to be as reliable or as conclusively authoritative as the administrators had supposed. Could Departments learn from one another how to manage a more open culture? Or should some types of decision be concentrated in a Department such as DOE which has more experience and more tradition of handling such issues?

The Precautionary Principle

The importance of taking proper account of public concerns and intuitions applies with equal force to the analysis of causes and effects. We know or think we know a good deal about some environmental causes and effects. On others, scientific knowledge or proof is incomplete. We cannot however afford to hold up all action until every fact is known and every cause and effect proved. The precautionary principle, informed by knowledge or public concerns, political perceptions and the wise judgment of decision makers ought to have a key part to play as well. Using the absence of conclusive scientific proof as an argument for inaction on an environmental problem that is causing popular and political concern can be dangerously short-sighted, and can lead to embarrassing political as well as environmental consequences.

This is the more important, because it is not infrequently the case that public perceptions and concerns, sometimes stirred up by well informed NGOs, turn out to be more reliable in the medium term than the cautious advice of government scientists mesmerised by the initial lack of conclusive scientific proof and the economic costs of precautionary action.

Acid rain is an example of an area where for several years the UK conspicuously failed to conduct policy in a broad enough way with enough regard for the precautionary principle. During the eighties we allowed policy to be too much determined by producer interests. We neglected to research sufficiently the effects of acid rain on humans, trees, buildings, etc., and refused to recognise the growing evidence from other countries. The upshot was that the UK became steadily isolated in Europe and at home. When belatedly the Government accepted the scientific case for dealing with acid rain, we were already a long way behind the general European consensus, and had to drag our feet lamely in pleading for more time for the necessary investments in cleaner technology, even though we were the largest contributor to the problem in Europe.

This case is a good illustration of the disadvantage of too narrow a perspective. Even our scientific base was too narrow at the outset, with too much weight being given to the scientific views of interested parties such as the coal industry and the energy producers, and not enough to more objective sources. The consciousness of the costs of adaptation loomed much too large in the minds of policy makers; and the real damage to the British and European environment, some of which has direct economic costs, was discounted because these costs and disbenefits did not show up clearly in the books of any one protagonist, and still less of the Government itself. Our concessions when they came had more of the character of giving way to European pressure than of any deliberate change of viewpoint.

I do not suggest that public opinion should be the sole guide to action, or that science should be ignored. But where there is serious public concern, and where the better NGOs are flagging the importance of an issue, it is always wise for Government and public bodies to pay the closest attention, to bring scientific resources to bear on the issue promptly (and from more than one camp if possible) and to consider precautionary action where appropriate. Governments ignore or discount public concern at their peril.

Economics and Policy Analysis

When we come to analysis of policy options and the work of the economists, even more difficulties arise with the rationalist approach. One principal problem is the establishment of appropriate values to attach to environmental benefits. But without this the whole project of cost benefit analysis as a guide to decision-making looks shaky. Much can be done on the proper assessment of the *costs* of policy options, and this should indeed be a discipline applied to all measures, whether in the environment or elsewhere; but when it comes to choosing which options give the most *benefit*, a whole range of factors need to enter the judgment. Quantified valuation techniques can sometimes be useful, but it is often necessary to go much broader into non-quantifiable political, social and distributional assessments of the impact of different measures. And it is vital not to let the decision-making process become the prisoner of what economists can quantify, and the conclusions of what such calculations appear to point to.

The successful international action to deal with damage to the stratospheric ozone layer done by CFCs and other ozone-depleting substances shows science and economics working well together for once to help shape a growing public concern and an emerging political strategy for action. Good science was, of course, essential for understanding the way in which CFCs and other substances damage the ozone layer in the polar regions; and good economics was needed to help weigh the costs of phasing out CFCs, and to help prioritise the pace of action on the different substances, and to allocate the burden of adaptation fairly. But neither science nor cost-benefit analysis determined the pace of advance. This had much more to do with the swelling tide of public concern and the political response to it.

Building Partnerships for Action

All effective measures in the environmental area need to have a broad base of political support, as well as a scientific and economic rationale. Otherwise they will not be adopted by the political process or will come unstuck in some other way. The debacle of the Major Government's attempt to increase VAT on domestic energy illustrates this very clearly. In this case there were excellent environmental/scientific and economic arguments for the change proposed, but their complete failure to develop in advance a coherent strategy for persuading the public and dealing with the social consequences undermined the proposal fatally from the outset.

This need to build effective political alliances for necessary policy changes will become increasingly important as the environmental agenda moves beyond traditional pollution control topics into the wider field of sustainable development, particularly sustainable transport, sustainable agriculture, and ultimately a sustainable economic policy. In all these areas the political concerns will be uppermost, and the environmental and sustainability agenda will have to contend with other strong agendas and political objectives. Good science and good economics will be essential elements of good sustainability policies, but they will be by no means sufficient. A broader basis of understanding and a stronger link between many different actors at different levels will be needed to make progress.

Air Quality: A Good Test for a Broader Approach

The current evolution of policy on air quality provides a good illustration of the kind of broad alliance between science, economics and the growth of public opinion and effective partnerships that will be necessary to build a sustainable policy for the future.

The battle for clean air has a long history, from the great Victorian struggles to control factory chimneys up to later campaigns for domestic smoke control. But in more recent years attention has gradually been shifting to emissions from vehicles. This is an area where there has been a satisfactory and symbiotic evolution of policy involving science, economics and technology working alongside and in parallel with the evolution and mobilisation of public opinion about health and quality of life as aspects of vehicle emissions.

A few years ago the Department of the Environment foresaw this growing concern and need for action and decided to take a number of related steps. They decided to improve and extend the monitoring of air quality throughout the country, and to operate it in real time so that air quality information could be made publicly available right up to the present moment. Linked with weather forecasting techniques it gave projections for the next day or two as well. By making this information widely available through the media, it was hoped to serve the dual purpose of informing the public (thus enabling individuals to take any immediate action appropriate to their own state of health and sensitivity to air pollutants) and building up public interest and demand for action over time to deal with some of the worst problems.

The second line of attack was to establish new high level scientific advisory committees on the impacts of different pollutants, and to strengthen the links with health research and the Department of Health on health effects. Here the Department wanted to respond to the growing public concern about air quality and respiratory disease, but at the same time to be sure that it was not simply stampeded into action on this or that pollutant, before the evidence (which is very complex in this area) was sufficiently clear for it to establish what kind of action would be most appropriate and effective.

At the same time as building up its knowledge about air quality and its effects the Department saw the need to help to develop a more sophisticated understanding among the public and in industry about possible measures and how they might be tackled. At one level this was a comparatively familiar though complex battleground: namely the various measures that can be taken to refine fuels more and to improve combustion in engines or elimination from exhaust streams or pollutants. The solutions to these problems are long term and expensive, but the oil and motor industries have the resources and the know-how to improve matters markedly over time, and the problem essentially was to marry the interaction of public concern and pressure, growing knowledge about the effects of air quality, and the handling of the big industries and their sponsoring departments to produce a progressive tightening of emission limits. This has not gone fast enough for some tastes, but it is a manageable and familiar process which, albeit with some hiccups, is producing results.

Emission controls may never, however, be sufficient by themselves to solve all air pollution problems; and they certainly will not be sufficient to deal with the related problems of CO_2 emissions from vehicles or of traffice congestion. Therefore we shall undoubtedly have to face up at some stage to the need to reduce growth in the volume of road transport. This brings us into really difficult political territory, well beyond the scope of normal bureaucratic processes to resolve. For here what are needed are behaviour changes freely entered into by the whole population. Economic factors and measures can play some part in influencing this, though we all know how insensitive to price changes the demand for car vehicle use is. More significant could be measures taken at local level to manage traffic, and to give preference to public service vehicles and pedestrians over unrestricted private cars. One has only to say this to see how the necessary measures could never be imposed by fiat alone, but must depend on vigorous public and local debate and informed assent or consent to the measures proposed, leading to agreement from everyone concerned that this will improve local environments.

The recent Air Quality Strategy issued by the Department of the Environment recognises all this explicitly, and reaches out to local authorities and all others concerned to see if there is a basis for building the necessary wide-based consensus for effective management of traffic and air quality. It is too soon yet to see whether this will prove to be an effective way forward. The key test will be whether it proves possible to build effective bridges between the national air quality strategy and the country's various transport strategies or plans. Only if such an integration can be made will we be on the way to sustainability in this area.

Conclusion

The rationalist approach to environmental policy-making based on scientific analysis and risk assessment is a tool, not a total solution. It needs supplementing and complementing with the views and opinions of the public and local communities in order to build a viable consensus for action, particularly as we move beyond the narrower pollution control agenda into the broader sustainable development agenda.

Those engaged with environmental policy-making, whether at international, national or local level, need to have a proper sense both of the potential and of the limitations of the traditional rationalist approach. Facts, science, economics and law are all essential aides to wise judgment, but they do not exhaust the matter and can never be reduced to a calculus. Non-traditional sources of information, careful listening to the voices of individuals and communities, and political considerations of the factors that underlie public confidence in environmental measures and Environment Departments need to play their part as much as the scientific analysis itself.

A wise politics in any area does not seek to impose solutions on the basis of one single model of reality. It seeks to build a consensus for action by reaching out to as many disciplines and modes of understanding as possible, and to as many different actors as possible at international, national and local levels, and among industries and non-governmental organisations, the media, individual citizens and communities. Not only will this build a broader basis of support for policies. It will also identify the part which all the other actors are prepared to play in a common endeavour to improve the environment, and thus build the basis for a much more successful overall politics for the environment than one relying on rationalist analysis alone.

Success breeds success. Failure breeds failure. The damage which a mistaken decision or policy or one which is shown to pay insufficient regard to the environment does to the credibility of a government can spread far beyond the particular issue in question, and have long-lasting effects. The period of intransigence of the UK Government during the late eighties on such matters as acid rain and sewage effluents in the North Sea did serious damage to Britain's environmental record and credibility, and made it much harder for the Government to make its voice heard as it deserved at home and abroad in the nineteen nineties. Brent Spar and the BSE debacle have had a similar debilitating effect.

In the British setting it used to be difficult to develop a long term policy stance on environmental issues at large. There has been a chronic tendency to deal with each subject one by one as it comes up, and to fall into the same kind of traps over and over again. This paper has argued that this is in part due to an inadequate but widespread paradigm within Government of what constitutes rational policy-making, with an excessive reliance on 'good' science and 'good' economics and an inadequate appreciation of other viewpoints and the concerns of the public and other non-governmental actors.

Two partial ways forward have been suggested. First to extend more widely the kind of open-ended consultative processes for policy-making and decision-taking that have been pioneered by local government in Local Agenda 21 and to some extent by the Department of the Environment on its better days. The culture of openness, consultation, partnership and of respect for other viewpoints is an essential part of making successful policies for sustainability.

Secondly we need to build up further the kind of broader-based strategies for sustainability that draw together many subjects and many concerns and integrate the contributions and viewpoints of many different actors. A broad sustainable development strategy embracing within it strategies for health and the environment, transport and the environment, energy and the environment and similar combinations, and drawing together the contributions and support of many different actors, is essential to building the broad basis of support needed. This will help avoid the kind of policy blunders that arise from thinking of issues one at a time from a narrowly rationalist point of view.

Biographical Note

Derek Osborn was Director General of Environmental Protection at the Department of the Environment, 1990–95. He is currently Chair of the European Environmental Agency as well as holding other posts and operating independently in environmental matters.

Local Agenda 21: The Renewal of Local Democracy?[1]

STEPHEN C. YOUNG

Local Government Interest in Sustainable Development and Local Agenda 21

Interest in sustainable development in British local government circles began in the late 1980s after the publication of the Brundtland Report in 1987.[2] The Local Government Management Board (LGMB) took the lead in promoting the concept on behalf of the local government associations. British local government went to the 1992 Rio Earth Summit with a detailed position statement, and members of the delegation were very active. The leading councils at this stage included Lancashire County Council, Kirklees, Peterborough, Leicester, Cardiff and Oxford.

In the period after Rio, interest in these issues in local government began to grow. The Rio accords had formally given all sub-national and local governments around the world the job of preparing Local Agenda 21 documents (LA21s) showing how they would address the whole range of issues dealt with in the global *Agenda 21* plan in their areas.[3] Interest continued to grow during 1994/5 after the publication of *The UK Strategy*.[4] The aim, as agreed at Rio, was to get all local authorities to prepare LA21s by the end of 1996, so they could be reported out to the UN Commission on Sustainable Development's review programme and conference in June 1997. However, LGMB fell back on aiming to get most councils started on their LA21s by the end of 1996. According to the 1996 LGMB Survey, 111 out of the UK total of 478 councils were aiming to complete their LA21s by the end of 1996.[5] Local government continues to be the main stronghold in Britain of serious interest in sustainable development, apart from the disciples working in and around the NGO community.

There have been three main reasons why different councillors have got involved with LA21 work. First, some undoubtedly moved into the whole policy arena of the environment as it expanded during the late 1980s and early 1990s simply because it gave them a role. From 1976 onwards they had faced round after round of spending cuts. During the Thatcher and Major eras they were confronted by centralisation processes, pressures to privatise services, and loss of powers. The environment was a policy arena where they could take the initiative. What emerged was a paradox—intense activity by a minority of councils in spite of the cuts and the controls.

Secondly, there were undoubtedly some individuals—councillors but mainly officers—in the pioneering authorities who were fascinated by the Brundtland challenge, by the problem of converting the principles underlying sustainable development into practical policies across the range of council responsibilities.[6] They were intellectually curious and understood the central point about trying to connect environment, society, and economy in new ways. They were trying to work out a more people-centred approach to planning and economic development. They became bureaucratic entrepreneurs, enthusing others around them, both before but especially after Rio.

Lastly, a number of external 'push' factors encouraged other councils to follow the pioneers. Before Rio there were influences like the Friends of the Earth Charter and the BT Environment Cities Programme. Especially after Rio, the LGMB played a strongly proselyting role. Within the professions, interest grew in planning and environmental health in particular. Increasingly in the mid-1990s networking followed on from academic writing and conferences.

The Range of Local Authority Activity

Many authorities started by analysing their own impact on the environment. Audits led on to energy-saving programmes, changes to purchasing policies, and a variety of other green housekeeping arrangements. Many councils have seen setting an example as important. Attempts have also been made to adapt decision-making processes. The argument here is that if sustainable development is to be promoted, its principles have to infuse all parts of a council's policy-making processes so that a holistic approach is established. Then all policy-makers can consider policy issues in terms of their effects on the environment. The principles underlying sustainable development then guide policy-making processes in the same routine way that the competitiveness initiative guides policy-making in Whitehall.

The adapting of decision-making processes is most in evidence in councils like Mendip and Stratford, where the environment unit was put into the Chief Exec's Department; or, as with Lancashire, where a strong and influential Planning/Environment Department was established. Committed political leadership, as in Kirklees, also drives change; but it is difficult to estimate how widespread these incidents are. Question 6 of the 1996 LGMB survey showed that 111 councils have established new officer liaison groups to promote LA21, while 113 are using existing officer working groups. In a small number of councils it seems that potentially significant changes are being established. But perhaps Manchester's experience is more typical. Attempts to promote holistic approaches broke down in a tangle of departmental and committee rivalries.

Local government and environment journals are full of policy initiatives to promote sustainable development. They range from air monitoring to re-using road surfacing materials; from Combined Heat and Power (CHP) projects to strategies to reduce car use; but these initiatives are patchy and limited. A lot of work has also been done on establishing baseline information via State of the Environment Audits and work on sustainability indicators. The LGMB survey (Q13) showed that sustainable development has had most influence on the traditional environmental areas like waste management, countryside issues, biodiversity, and land-use planning. It has had least impact on social services, welfare, anti-poverty strategies, housing, tourism, economic development, investment strategies and tendering.

There is a considerable contrast with the experience of Western Europe. On the continent, cities, communes and other sub-national councils are focusing more on policy integration, holistic approaches and technical solutions. However, somewhat surprisingly, they see Britain as a world leader on the participation front. This reflects the fact that in Britain the environment agenda has broadened out at the local level to become one principally concerned with community involvement and democratic participation.

LA21 Participation Strategies and Their Impact

One of the most widely used distinctions in discussions about participation is the difference between top-down and bottom-up approaches.[7] The top-down approach is a one-way consultation process dominated by the council, and used mainly to pass out information about what it is doing. The council controls the agenda and there is little scope to change its position. In contrast, the bottom-up approach is a two-way process that develops into a genuine dialogue between the council and local communities based on an open agenda and a sharing of information. Policies and priorities are open to discussion and the aim is for local people to own the participation process.

The features of the LA21 participation programmes that stand out are variety and imagination. Traditional approaches based around leaflets, exhibitions and public meetings have been submerged in a riot of experimentation. They involve new ways of sharing information; local residents designing questionnaires and doing house-to-house interviews; arts-based approaches; and other attention-grabbing events. The central aim is to draw in the views of all local 'stakeholders'.

Visioning techniques are used to get people to discuss how they would like a neighbourhood to be, prior to discussing the action needed to change it. Community profiling and village appraisals involve local people doing surveys to identify local needs and available resources, and priorities for change. Small group discussions and working groups are used to get people together for a meeting or to divide up a large meeting. Visioning and profiling techniques are then used to identify key issues. Focus groups draw in individuals from specific backgrounds to discuss pre-set agendas. 'Planning for Real' exercises bring local people together for short intensive periods to promote consensus about the nature of neighbourhood change. Environment Forums, Round Tables and a range of advisory committees are used to get stakeholders together to generate discussion, commitment and consensus.

Probably about 50 or 60 councils have aimed at something like a bottom-up strategy. These include Kirklees, Reading, Gloucestershire, Derbyshire, Vale Royal, Nottinghamshire, Mendip, Croydon, Merton, and Leicester. Others have tried similar approaches in the context of the Single Regeneration Budget (SRB) and estate regeneration programmes. In practice, however, many councils have fallen back on a 'Yes . . . But . . .' strategy. This occurs where the council adopts the rhetoric of the bottom-up strategy, but finds it difficult to let go of the agenda. What happens gets summed up by a senior figure as follows: 'Yes let's aim at a Bottom-Up Strategy, but the issue of the ring road/opencast mine/landfill site is too important to compromise on.' The council thus changes the nature of its approach as the participation programme develops. It evolves into a more limited and controlled exercise.

The widespread interest in participation has certainly had an impact on participation processes. Councils adopting the more innovative approaches have had positive responses from a wider range of people getting involved than the usual self-presenting groups. However, by the end of 1996 very few LA21s had actually been completed. It was thus too soon to assess fully the impact of the participation programmes. Although great claims were being made, we were probably at that point still not yet at the peak of what might be termed the 'enthusiasm stage'.

It is always necessary when assessing participation to identify the perspective a programme is being judged from. Thus a council adopting a top-down strategy would judge 'success' in terms of its out-of-town industrial estate proposal not being irreversibly damaged. On the other hand, a council pushing a bottom-up strategy would focus on the extent to which it was empowering local communities and putting their views at the heart of policy-making processes. In the latter case, it is only possible to judge success in the longer term. It is too soon as yet to assess the impact of the LA21 participation programme on the LA21 document itself; or that document's influence on the revision of a council's statutory plan and the restructuring of its budget in ways that reflect sustainable development principles.

LA21 participation programmes are encouraging the growth of partnerships between local communities and other agencies.[8] These were picked out at Rio as having a significant contribution to make to the implementation of LA21s. These community-based partnerships operate in what is increasingly referred to as the social economy between the public and private sectors. Examples include recycling schemes, housing co-ops, Local Exchange Trading Systems (LETS), credit unions, environmental improvement schemes, wildlife organisations, and a variety of local development trusts.

These community-based partnerships take many forms but have three main features in common. First they operate on a not-for-profit basis. Secondly, they focus on the level of the local community. This is usually meant in the geographical sense of the village, the estate, the urban neighbourhood. But it also refers to communities of need, interest and experience across a wider area, as with ethnic groups across a city. Finally, these organisations emphasise local democracy and the involvement of local people in defining their needs, shaping programmes, and controlling the development of the organisation.

A number of difficulties have emerged with LA21 participation programmes. People have been asked about the promotion of sustainable development and priorities in a situation where the extent of future resources is not clear. For example, both a shift to energy supplies coming from more environmentally friendly sources and to better and more extensive public transport require substantial capital investment. With regard to urban traffic congestion, participation programmes have produced extensive calls for change; but it is very difficult for councils to respond positively because of the government's lack of response to the urban traffic agenda set out in the Royal Commission on Environmental Pollution's 1994 report.[9] In addition, privatisation has led to the fragmentation of public transport undertakings and made it much more difficult to promote integrated investment programmes.

Some councils have found it difficult to maintain momentum. Some of the early enthusiasm for the Environment Forums, for example, has ebbed away amidst criticisms that they have become fossilised talking shops. In some places it has proved difficult to draw lower income groups and young people into the participation programmes. The problems of raised expectations haunts some councils. The businesses that have got involved tend to be committed to sustainable development anyway. The uninterested firms that have ignored the calls to give their views are the ones it is important to influence. But organisations like chambers of commerce—and in farming areas the National Farmers' Union—can only speak for their members. They cannot commit them to using less fertiliser or transporting more goods by rail.

The other main problem is that it has proved difficult to integrate the experiments with participation with the need for strategic planning. Many of the pioneering councils' participation programmes have focused on the micro-level, the level of the village and the urban neighbourhood. However, work on LA21s has also clarified the need for a strategic level of analysis. On waste, for example, a council needs to decide the balance between landfill, recycling, and incineration (including the CHP option) within its programme for the next decade. Promoting recycling, reuse and waste minimisation can help; but this does not remove the need for new landfill sites. Transport also has to be planned at the strategic level. Similarly a city-wide or county-wide approach has to be taken to identifying land that needs to be released for housing or industry or a new site for a major football club. The use of vacant urban sites can help reduce, but not remove, the demand for land release. In practice many LA21 participation programmes have been taking place in something of a vacuum without the implications of strategic issues like these being fully integrated.

The strategic approaches and the micro-level approaches have been running in parallel. This sometimes makes it difficult to reach consensus as the two have to be reconciled and the strategic approach may well limit options at the neighbourhood level. What is needed is more of a linear approach to participation over years not months. If people are presented with proposals for a new reservoir or landfill site on their doorstep they are understandably horrified. But if they are involved via citizens juries and similar arrangements, then they can understand the technical details that lie behind the proposals being made, and see if different conclusions can be drawn. Such an approach is used with regard to water planning in the US; and work is being done on it in the context of waste management in Britain.[10]

Explaining the Interest in the Participation Dimensions

A number of factors contributed to this growth of interest in the participation aspects of LA21. First of all, Chapter 28 of Agenda 21, as agreed at Rio, stresses the need to involve all groups in society when preparing LA21s. Particular mention was made of drawing in women and young people; involving all groups in society and not just the more articulate ones; and of trying to generate consensus. Participation was thus promoted to being an integral part of the LA21 policy-making process. In the LA21 context it was no longer—as so often in contemporary politics—an optional extra. Rio thus elevated community involvement to a new status.

Interest in participation was further prompted by two other factors. The LGMB has stressed the importance of promoting community involvement in its publications about LA21 and its monthly mailings to environmental co-ordinators. This has encouraged individual councils to give it more attention and be more innovative.

In addition, those involved within individual councils in promoting the participation aspect of LA21 have brought new approaches to the task. They have come in with experience of community development, running adult education programmes and directly involving people in countryside and recreational projects, as via the Groundwork Trusts. There has also been an influx of people from the NGO world with detailed ideas about how to promote participation. Collectively the attitudes of all these people towards participation reflects much more of a bottom-up, people-centred perspective than that of planners who have previously been running top-down and use plan consultation programmes. The views and values of these people have decisively influenced the nature of the participation programmes followed by the more innovative councils.

There has been a broader change of attitude towards community involvement amongst some policy-makers. The resulting exploration of new ideas has not just been directed at LA21. It has also found expression in the contexts of housing renewal and estate regeneration; City Challenge; and the SRB programmes. The attempts to promote bottom-up style strategies reflect a growing recognition of the need to move on from the top-down, bureaucratic, paternalist approaches that predominated during the 1960s and 1970s. The process of imposing projects with minimal consultation programmes created schemes that did not work, such as high-rise flats and the Hulme crescents. A range of subsequent inner city regeneration programmes did promote economic renewal and make a physical impact in terms of new buildings and landscaping projects. But they did not produce social renewal and tackle local unemployment, housing deprivation, crime and other dimensions of urban poverty.

As a result, a small but growing number of policy-makers began to argue the need for a new approach. Their starting point was that top-down solutions do not relate effectively to peoples' perceptions of what is wrong. They argued that the lesson from previous decades was that it was important to start with a blank agenda, and ask local people to define the problems as they perceived them. It then became possible to produce relevant and creative solutions. These have more chance of working, because they draw from local peoples' knowledge and experience. The solutions that emerge are promoted and supported by local people. They are not just more acceptable to them: they are also more likely to be protected by local people. These arguments are reflected in the discourse of City Challenge and the SRB in the early 1990s about the importance of drawing in all the stakeholders, and empowering local communities.

This new emphasis on bottom-up approaches has also found support from another constituency. The early 1990s produced a debate in the media and amongst academics about social exclusion, alienation, a loss of trust in local government, and a loss of faith in the capacity of local political institutions and processes.[11] Despite the attempts to communicate more effectively with local people and the experiments with decentralisation and neighbourhood offices, many were increasingly concerned about the growing disillusionment with local democracy. Discussion amongst media commentators, academics, politicians and leading local government figures focused around the need to regenerate local democracy. These people were not specifically interested in the environment or LA21. They were drawn to the bottom-up approaches to participation from a different perspective. They were interested in the experiments because of the opportunity to try new ways of communicating with the public and promoting community involvement. These new approaches thus offered a way of breathing new life and vitality into the institutions and processes of local government itself.

Empowerment fits into this broader context because of its links to the citizenship debates. It promotes conceptions of citizenship in the civic republican tradition. These build from a set of rights to emphasise participation, the practice of citizenship, and involvement in debate and decision-making. This runs counter to the Right's more liberal interpretations which were prevalent in the 1980s and early 1990s. They emphasised citizens being able to make more choices as parents, tenants, consumers, and customers of health services.

A Shift In Local Governance?

The most important part of LA21 has been the way in which participation programmes have become a conduit for the unleashing of energy and ideas into the arena of local democracy. These approaches are different in form from any others adopted since the 1969 Skeffington Report.[12] They represent British local government's most serious attempt to date to empower local communities and to give them greater influence over what happens around where they live.

These participation experiments have been extended and reinforced by the promotion of community-based partnerships in the social economy. LA21 participation programmes are creating opportunities to identify potential projects. The significant point here is that the establishment of these not-for-profit organisations represents a practical response to local need as a result of visioning exercises. The importance of these organisations is that they offer a way of linking sustainable development, community participatory democracy, and place, at the micro level of the village, the estate, and the urban neighbourhood.

These partnership organisations in the social economy are growing in numbers to the point where they are becoming a 'Third Force' outside the state and market sectors. The contrast with firms is important here. Capital is free to move, but these community-based partnerships are closely linked to place, to improving conditions where their members and stakeholders live. The growth of these organisations also fits in with the wider changes in local governance, with the continued moves away from bureaucratic paternalist welfare provision towards a different model. Increasingly, smaller, more flexible, enabling councils are working, not just through partnerships with industry, but also with a whole variety of voluntary and not-for-profit organisations. By helping to promote empowerment and the expansion of the social economy, LA21 is helping to shift the balance of provision between public, private and voluntary sector bodies. If this shift in local governance is to take place on a significant scale, central and local government need to develop capacity-building programmes to target skills and resources more effectively, following the example of councils like Barnsley. This will help to build up social capital and strengthen the weakened links between the local state and civil society.

The experience of the more radical LA21 councils, and of some of those from the SRB and other renewal programmes, in promoting participation and community involvement suggests that it is at the micro-level where people are most responsive. It is at the level of the urban neighbourhood, the estate, and the village that bottom-up empowerment programmes can be developed as a means of tackling social exclusion, alienation, and loss of faith in the capacity of local government to deliver. This is the level where empowerment has the greatest potential and where it can be linked to the development of the social economy.

145

The main problem with promoting such approaches is that it complicates the role of the elected councillor. At present, British local government is established on the basis of the representative democracy model. Ultimately, after all the participation programmes have been completed, local governments have to govern. When there are conflicts at the micro level, or between strategic and more local approaches, councillors have to make decisions in the public interest for the whole of their areas. However, some councillors feel that the experiments with participation are starting to undermine this role, and to leave too many decisions about local priorities to local people. In addition, some of the bigger community-based partnerships are making decisions about resource allocation at the ward level. These developments are starting to move us towards more of a participatory democracy model, as is the use of IT.

The way forward here is to focus on re-engagement with local communities at the micro-level; and to do more research, both on the impact over time of neighbourhood-based participation initiatives on council programmes, and on ways of integrating the strategic and neighbourhood levels. This will help us to work out how councillors can redefine their role, while at the same time helping us to regenerate local democracy.

There is a tremendous opportunity here for the innovative, people-centred approaches to participation of the mid-1990s to become established as best practice. This is what happened to some of the ideas pushed by the New Urban Left councils in the 1980s. Their policies towards the disabled and ethnic minorities—derided by some at the time—have been widely adopted as the mainstream, conventional approach a decade later. However, reaching the watershed on participation that is now coming into view ahead would have a much broader significance for the regeneration of local democracy.

Biographical Note

Stephen Young is Senior Lecturer in Government at Manchester University; and co-editor of *Environmental Politics*. Publications include *The Politics of the Environment*, 1993; and with G. Stoker, *Cities in the 1990s*, 1993.

Notes

1 The author is grateful for financial support for the research on which this piece is based from the ESRC's Local Governance Initiative—Grant Ref L311253061.
2 World Commission on Environment and Development, *Our Common Future*, Oxford, OUP, 1987.
3 M. Grubb, *The Earth Summit Agreements*, London, Royal Institute for International Affairs, 1993.
4 Department of the Environment, *Sustainable Development: The UK Strategy*, London, HMSO, Cm. 2426, 1994.
5 B. Tuxworth and E. Thomas, *Local Agenda 21 Survey, 1996*, Luton, LGMB, 1996.

6 See for example A. Blowers, ed., *Planning for a Sustainable Environment*, London, Earthscan, 1993; G. Haughton and C. Hunter, *Sustainable Cities*, London, Jessica Kingsley, 1994; M. Jacobs, *The Politics of the Real World*, London, Earthscan, 1996; P. Selman, *Local Sustainability*, London, Paul Chapman, 1996; and J. Kirby, P. O'Keefe and L. Timberlake, *The Earthscan Reader in Sustainable Development*, London, Earthscan, 1996.

7 This section draws from S. C. Young, *Promoting Participation and Community-Based Partnerships in the Context of Local Agenda 21: A Report for Practitioners*, Manchester, EPRU Paper, Government Department, Manchester University, Manchester, M13 9PL; D. Wilcox, *The Guide To Effective Participation*, Brighton, Partnership Books, 1994; Department of the Environment, *Community Involvement in Planning and Development Processes*, London, HMSO, 1995. These three have masses of further references.

8 S. C. Young, *ibid.*; S. Thake, *Staying the Course: The Role and Structure of Community Regeneration Organisations*, York, Joseph Rowntree Trust, 1995.

9 Royal Commission on Environmental Pollution, *18th Report: Transport and the Environment*, Cm 2674, London, HMSO, 1994.

10 S. G. Walesh, 'Interaction with the Public and Government Officials in Urban Water Planning', in H. van Engen, D. Kampe, and S. Tjallingii, eds., *Hydropolis: The Role of Water in Urban Planning*, Leiden, The Netherlands, Backhuys, 1995.

11 See R. Wheeler, ed., *Local Government Policy-Making*, March 1996 (special issue on Empowerment and Citizenship); P. Macnaghten, R. Grove-White, M. Jacobs, and B. Wynne, *Public Perceptions and Sustainability in Lancashire: Indicators, Institutions, and Participation*, Preston, Lancashire County Council, 1995; and Commission for Local Democracy, *Taking Charge: The Rebirth of Local Democracy*, London, the Commission, 1995.

12 J. Gyford, *Citizens, Consumers and Councils*, London, Macmillan, 1991.

Voluntary Associations and the Sustainable Society[1]

BRONISLAW SZERSZYNSKI

As this is being written, in early 1997, it is almost five years since the United Nations Conference on Environment and Development (UNCED) brought an unprecedented number of political leaders together at Rio de Janeiro to discuss how the world might move towards sustainable development, towards economic and social development that, in the words of the Brundt-land Report, 'meets the needs of the present without compromising the ability of future generations to meet their own needs'.[2] To mark this anniversary, in June of this year the United Nations General Assembly will be holding a Special Session in order to review the progress to date made by more than 150 nations towards meeting the commitments they made in 1992.

However, if the British case is anything to go by, the Special Session is not likely to be marked by a mood of optimism. Transport trends alone in Britain indicate just how far we are away from achieving sustainable development. The Department of Transport's National Road Traffic Forecasts suggest that traffic is set almost to double over the next thirty years, and to grow even more in rural areas.[3] However sceptical one might be about the robustness of such forecasts, they nevertheless give a stark indication of the difficulty of achieving anything like environmental sustainability.

One important reason for this lack of progress has been the failure to secure widespread public identification with and participation in sustainable development initiatives. At UNCED it was recognised that public participation would have to be central to the pursuit of sustainable development, and this principle was enshrined in Agenda 21, the action plan agreed at the conference. There were a number of reasons for this participatory emphasis, but perhaps the most important was the sense that the changes that would be required of societies would be so widespread, and so deep, that the active involvement of as many sectors and as many individuals as possible would have to be secured.

It was informed by this recognition that the UK's sustainable development plan, published in 1994, announced the creation of a Citizen's Environment Initiative, whose goal was 'to increase people's awareness of the part that their personal choices can play in delivering sustainable development, and to enlist their support and commitment'. Launched in 1995 under the title of Going for Green, this initiative has attempted, using methods such as leaflets, mass media advertising and resource materials for schools, to motivate the public to make lifestyle choices congruent with the goals of sustainable development. At the local level the emphasis on public participation has

© The Political Quarterly Publishing Co. Ltd. 1997
Published by Blackwell Publishers, 108 Cowley Road, Oxford OX4 1JF, UK and 350 Main Street, Malden, MA 02148, USA

been perhaps even stronger, with the Local Government Management Board advising local authorities to make awareness-raising activities and broad consultation processes key elements of their sustainable development strategies. However, despite its vigorous progress in certain institutional circles such as local government, sustainable development is far from achieving widespread public support. The evidence from MORI surveys, for example, is that after an initial surge at the end of the 1980s, public commitment to making lifestyle choices in defence of the environment has remained at low and at best constant levels (see Worcester, this volume). Most of the population have their own serious anxieties about environmental change, but find 'sustainable development' a remote and alien concept. Most, similarly, might acknowledge in principle the need for people to make changes in their lives, but feel such a degree of scepticism that their solitary action could make any difference that they dismiss the idea of lifestyle change as pointless self-denial.[4]

How can greater public identification with the aims and values of sustainable development be secured? How might we overcome the cultural barriers that might be preventing people from acting in concert in defence of the natural environment and the viability of their own communities? How might society be organised to reverse the tendency for the individual pursuit of happiness to be in competition with the common good? I want to offer in this chapter a number of different reasons why the valuing and fostering of voluntary associations might come to play a significant role in the pursuit of sustainable development. I suggest that such associations—all the way from formally constituted charities and friendly societies right down to the most *ad hoc* and informal of alliances between people—can play a crucial role in the generation and implementation of sustainable development initiatives; that high levels of associational activity in society can help foster the cultural conditions of trust and public-mindedness in which individual and institutional action in defence of the environment can better flourish; and also that associational activity in itself can serve as the vehicle for the sort of human flourishing and quality of life which is a key goal of sustainable development.

Associations and the Implementation of Sustainability

Stephen Young's contribution to this collection provides us with a useful survey of ways in which groups and organisations in the 'social economy' or 'third sector' can play a significant part in the implementation of Agenda 21 at the local level. Local environmental, wildlife and amenity groups already, of course, play a critical role in the management and improvement of open spaces. Voluntary associations, both formal and informal, are vital to the functioning of many recycling and car-sharing schemes; a network of permaculturists, allotment holders and organic gardeners form an infrastructure for the generation, preservation and transmission of ideas both old and new in ecological food production. Food purchasing co-ops and organic box schemes help provide the support structure for farmers and growers moving to more environment-friendly methods of growing; and, of course, spontaneous protest groupings help both in the recognition of local environmental problems and in the stimulation of the national and wider political debate.

Of course, sustainable development is not just about 'environment' in the strict sense, serving as it does as a vehicle for a huge range of aspirations for a better world. As a political concept it has thus expanded outward from the strictly environmental to embrace qualify-of-life and equity issues as diverse as global trade, education, welfare and local democracy. This expansion is grounded both in the ethical acknowledgement that environmental aims should not be considered in isolation from issues of equity and justice, and also in a more prudential recognition that inequity and poverty themselves contribute to social and environmental unsustainability, and thus need to be attended to by environmentalists.

The capacity for voluntary associations to help realise sustainable development in this wider sense is if anything greater than in the more narrowly environmental area. Self-help groupings such as LETS (local exchange and trading) schemes, credit unions, housing co-operatives, tenant co-operatives, alternative health projects, sports and leisure clubs, child-care initiatives and community arts projects may not have explicitly environmental foci, and in most cases may have arisen quite independently of any government initiative, but nevertheless are *de facto* manifestations of the broader agenda of sustainable development.[5] Such voluntary associations display extraordinary inventiveness and energy in the pursuit of quality-of-life goals—both in the meeting of existing human needs and also in the creation of new possibilities for a fuller human flourishing beyond simple subsistence.

It is important to distinguish these kinds of self-generating initiatives from consultative exercises such as those carried out by local authorities under Agenda 21. Such exercises, however genuine the intention, are circumscribed in a number of ways. They are always in some sense on the terms of the body which is doing the consulting; they are circumscribed in time, in that they have a clear beginning and end; they are restricted in terms of the kinds of activity members of the public can engage in during the exercise; and participants, it has to be said, generally feel they get very little out of participating in them, in terms of influencing outcomes or of any more indirect, personal benefits of participation. Voluntary associations, by contrast, are spontaneous constellations whose remit is determined solely by their participants; their lifespan is determined by the members themselves; they can involve an extraordinary range of different kinds of activities, as a glance at any CVS Directory of Local Voluntary Organisations would indicate; and, judged on their persistence, variety and vitality alone, they clearly offer a range of satisfactions to their members.

There are many different reasons why it might be a good idea to devolve as many sustainable development activities from central and local government to such associations.[6] Some, more negative in character, concern the capacity for large bureaucratic organisations to deliver the goods. The centralised state, despite two decades of neo-liberalism, can still be seen as struggling to fulfil all the functions it has taken to itself in this century. This is partly due to the sheer spread of its activities, and partly due to the huge social transformations that have taken place since the early 1970s. Not only does the globalisation of economic activity and the rise of regional government mean that nation states are far less capable of controlling what happens within their own boundaries than they might have been in the immediate post-war era; but also the fragmentation of values and lifestyles within society has meant that the top-down bureaucratic model of regulation and provision is increasingly ill-matched to social realities. The area of health is a good example here: the complexity and sheer variety of ways in which people have come to think about their bodily, mental and emotional health in recent decades is making any top-down notion of meeting 'health needs' increasingly hard to sustain. As Paul Hirst has written, not only does 'bureaucratic monoculture' threaten to stifle human autonomy and creativity, it also threatens to produce a no less dangerous response of refusal, as more and more people may react to it by disengaging from any sense of a public world and retreating into purely private concerns.

Another more positive cluster of reasons to look to associations for the pursuit of sustainability relate to their own particular characteristics and strengths. Because of its natural rootedness in public concerns and enthusiasms, associational activity is much better placed than are bureaucratic modes of social organisation to generate environmental and quality of life objectives which will align with people's own everyday experiences and concerns.[7] While locally based voluntary associations alone may not be capable of responding adequately to environmental problems of the largest scale, they may nevertheless be better placed to identify targets and priorities in respect of local problems and concerns, and even to generate appropriate local responses to transnational problems such as global warming.

A related advantage of voluntary associations over bureaucracies in this respect is the way that their agendas and activities are more likely to be 'owned' by their participants, as opposed to being felt to have been determined and imposed from outside. The social authority of bureaucratic institutions cannot be dismissed lightly, but it is a distinctive kind of authority, grounded in ideas of impartiality, rationality, efficiency and expertise. Initiatives such as those pertaining to sustainable development which originate in such institutions tend to carry this flavour with them, so that attempts they might make to enrol others can be felt as disciplinary and alien. Voluntary associations can, of course, manifest many of the positive features of bureaucracies, but their impressive power to evince loyalty and commitment amongst their participants seems to derive from a very different kind of social authority, one based on their status as authentic and spontaneous expressions of public will. A kind of sustainable development that was rooted more firmly in the principles of voluntary association would be far more likely to be congruent with the patterns of everyday life, to be experienced as working with rather than against the grain of people's needs and inclinations, and thus to attract more of the seemingly boundless social energies that animate associational activity in modern societies like Britain.[8]

Associations and the Civic Virtues

There is a second broad set of reasons why the extent of associational activity in society might be critical to the delivery of environmental sustainability. These relate not so much to the specific activities which might be done by voluntary associations, but to the longer-term moral and cultural consequences of associational activity in general. My argument here is that the achievement of any kind of sustainability will depend as much on cultural factors as on technological and institutional ones, that associational activity in society is central to such cultural factors, and thus that generating and defending the conditions in which voluntary associations can flourish has to be central to providing the kind of cultural infrastructure that a sustainable society will need.

The idea that sustainable development might require not just intergovernmental agreements and commitments but also a far deeper and broader set of cultural changes in wider society is, of course, not new. The authors of the 1972 Club of Rome report insisted on the necessity of 'a basic change of values and goals at individual, national, and world levels' if the world was not to run catastrophically up against natural limits; and the Brundtland Commission, in turn, acknowledged that they were in effect calling for 'changes in human attitudes' which would require 'a vast campaign of education, debate and public participation'.[9] However, such calls for a change in values have generally focused on those which are directly environmental, rather than recognising the effect that more general human sensibilities can have on the sustainability or otherwise of a given society. What I am suggesting here is that the possibility of achieving sustainability might depend as much on what sort of people we are as what sort of things we do, and thus that we have to pay attention to how our moral characters are shaped by the kind of institutions, both formal and informal, that our lives are caught up in.

Questions about moral character have been high on the public agenda in recent months in Britain and elsewhere. Much of this concern has taken the form of anxiety about a widely observed slow trend towards what might be called 'hedonistic individualism'. This kind of moral orientation, seen as at once a cause and a symptom of declining social cohesion, has been discussed under two aspects. First, writers have expressed concern about a seeming diminishment of moral empathy, a focusing of energies on to the improvement of one's own life rather than that of others and of society as a whole. Under such conditions, even an expressed concern for family can appear more like extended egoism than genuine altruism. Secondly, moral commentators have lamented a narrowing of the criteria of the good life by which

152

people's lives are governed, a shrinking away from the struggles, both private and shared, which might be associated with genuine self-development, and a gravitation towards the easier pleasures available in a society organised around consumption. Drawing, however distantly, on virtue theorists such as Alasdair MacIntyre, such debates have had the effect of promoting the idea that human moral experience is not just characterised by the application of abstract moral principles at a succession of discrete moral dilemmas, but also by enduring dispositions, by habitual attractions and aversions which may—or may not—bend us towards the good in given situations. These debates have thus had the effect of focusing attention on how moral character might be shaped by the different social milieus through which people pass during their lives.

However, in this public debate the language of 'virtue' and 'character' has tended to be the preserve of the cultural conservatives, and this in a double sense. First, the virtues identified as those which should be prized and cultivated have tended to be those that would underwrite rather than challenge the status quo. This moral conservatism should be of concern to environmentalists as much as to feminists, since one might legitimately wonder whether a move towards a sustainable society would not require virtues that clash dramatically with those conducive to the current social order. But, secondly, writers such as Amitai Etzioni who are keen to emphasise virtue and character exhibit a strong bias towards those social relationships into which we are thrown rather than choose to go—those of family, geographical community, religious membership, and so on, rather than the voluntary associations discussed here. The implication is that it is through a fuller participation in Hirst's 'communities of fate', rather than his 'communities of choice', that character can be shaped in a way that leads us away from rather than towards a culture of hedonistic individualism.

However, recent research suggests that societies which are structured less by traditional memberships and obligations and more by individual choice can actually exhibit more, rather than less, other-regarding behaviour. First, some researchers have pointed out that voluntary associations are often saturated with particular moral norms and character ideals, which, once one has entered such a collectivity, guide one towards making particular *kinds* of choice. Thus certain specific associational contexts might well be conducive to moral dispositions which would be needed for a transformation towards sustainability. For example, Helmuth Berking has been exploring the curious fact that modern, individualised societies such as America and Germany are characterised by vast amounts of other-regarding behaviour, such as the giving of time and money not just for proximate but also for distant, anonymous others. Arguing that this is not just a survival from earlier, more collectively oriented times, he suggests that it is the product of a distinctive kind of solidary individualism generated by modern cultural conditions. A central role in this process is played, for Berking, by voluntarily-entered 'lifestyle coalitions', which provide, through their substantive cultural content, the character-ideals which propel the otherwise utilitarian individual towards solidary relations. It is perhaps in these cultural milieus—generally not those valued by moral commentators—where the virtues of any emergent global citizenship might be forged.[10]

Secondly, apart from the specific virtues that might be encouraged by particular *kinds* of voluntary associations, there are other more general dispositions, nurtured by the very process of associational activity itself, that would be vital for any sustainable society. In an important book, Robert Putnam has argued that the differing fortunes of the North and South of Italy, in respect of the effectiveness of their regional governments, their economic performance and the happiness of their citizens, is best explained, not by levels of industrialisation and economic modernisation, but by the deeply rooted traditions of civic association in Northern Italy, he suggests, has long been characterised by 'an active, public spirited citizenry, by egalitarian political relations, by a social fabric of trust and co-operation', while the South is 'cursed with vertically structured politics, a social life of fragmentation and isolation, and a culture of distrust'. It is this deep-rooted cultural difference, itself perpetuated by starkly different levels of associational activity—of what Putnam calls 'social capital'—that is at the root of the different performances of the North and South.[11]

Trust

How does this relate to sustainability? The important thing to note here is that, for *this* part of the argument, it matters less what people are doing together than whether they are doing something at all. Associational activity, even where it is not actually directed toward environmental goals, helps to create the cultural preconditions for sustainability by generating and sustaining dense horizontal bonds of trust and co-operation. It thus creates a generalised climate of trust where people are less likely to think simply of their own, narrowly-conceived interests.

This is crucial if anything like sustainability is to be achieved. Most environmental goods, such as clean air and biodiversity, are public goods, in the sense that (a) they are indivisible, and (b) when they are consumed they are done so collectively. Such public goods are the first casualties of individualisation, according to the well-known logic of the 'tragedy of the commons'. If one cheats by taking more than one's fair share of a public good—such as by emitting a great deal of pollution into the atmosphere—the costs are shared but the benefits accrue to oneself alone. Unless one can trust others to be, in general, as public-spirited as oneself, it is only by an act of unusual self-denial that one will resist taking more of, and thus eroding, a public good. If one has no confidence that anyone else is reducing use of the car or of domestic energy, then to do so oneself will feel like pointless self-denial.

But to speak of trust is not simply to call forlornly for trustworthiness. To ask here whether the confidence that a given individual might have in others is well-founded or not would rather miss the point. Generalised trust of others, just like generalised distrust, is self-fulfilling. For example, if everyone behaves *as if* others are generally untrustworthy, then people will actually be so. What matters most at the level of society as a whole is less whether people can make well-founded judgments about others, than whether people are of such a character as to be inclined towards trust and co-operation. From this perspective, trust is less a matter of individual discernment and cognition and more social glue that binds people into relations of mutuality. However, under cultural conditions where habits of trust and co-operation have atrophied, forms of action which depend on trust of others, such as those in defence of public goods, environmental or otherwise, will be almost impossible to sustain. This kind of cultural dynamic contributes to the present situation in transport, where every car driver wishes others would use other forms of transport to leave the roads clearer for him, but few actually decide to do so, resulting in spiralling congestion and a declining quality of life for all.

Associational activity can thus contribute towards environmental sustainability by encouraging the cultural conditions of trust and public-spiritedness which are conducive to the defence of public goods. Jürgen Habermas has argued that voluntary associations such as the new social movements are vital if society is to make appropriate political decisions about issues such as the environment. For Habermas such movements play the crucial role of defending 'communicative rationality' against the encroachment of narrow instrumentalism, of providing micro-public spheres within which the formation and refinement of shared judgments about public issues can take place. I am making a further suggestion here, that associational activity is conducive to the defence of public goods, not just because it makes possible the practice of shared, deliberate judgment, but also because it encourages moral habits of co-operation and public-spiritedness in its participants.

There is another important way in which the achievement of environmental sustainability might depend on the general level of associational activity in society. Putnam used his Italian data to argue that horizontal trust, that between citizen and citizen, also facilitated what might be called vertical trust, between citizen and authority. Contrary, perhaps, to the expectations of many contemporary moral pundits, Putnam found that the regions of Italy which were least civic and most characterised by individualism and distrust were those of the South, dominated by a traditionally-based culture centred on personalistic ties of family and local community. The most civic regions were in the more urban, highly educated North, characterised by traditions of voluntary association rather than by customary bonds and loyalties. In the uncivic regions any ties that there were between the citizenry and those in authority were not the generalised bonds that exists between those engaged in a shared pursuit of the common good, but a bond that was personalised and instrumental in nature, one based on what each side could get out of the relationship. Although they were cynical about those in authority, people in un-civic areas had little sense that they should or could do anything about it; it was not just the holders of office, but also the office itself which was regarded with disdain. Interestingly, it was from the regions of the South, where trust in one's neighbour was lacking, and where everyone expected everyone else—including those in authority—to be corrupt, that the call for strong government was heard most frequently. In the North, the greater degree of trust in others and internalised norms of co-operation meant that government was able to go with the grain of culture, and thus operate effectively with a lighter touch.

Such observations are crucially relevant to the pursuit of sustainable development. Indeed, Putnam's characterisation of the Italian South seems remarkably applicable to the British polity in the area of environment and technological change, characterised as it is by a deep and qualitatively new cynicism about political institutions, scientific experts and so on. Research suggests that, although people broadly approve of the values and priorities of sustainable development, they remain deeply suspicious of official programmes in this area, due to lack of trust in the relevant institutions. Initiatives such as Going for Green which seek to encourage people to make environmentally positive changes in their consumption and wider lifestyle are fatally undermined by such lack of trust. The response of the public to the introduction of new technologies such as nuclear power and biotechnology seems to be shaped as much by their sense of trust in the institutional framework in which the technology is embedded as in the 'objective' riskiness of the technology itself. Public trust and distrust of authority is one of the key issues affecting society's ability to make appropriate choices in relation to our technological and environmental future.[12]

Putnam's work shows us that if we want to know why a citizenry's attitude towards its public figures and institutions is one of cynicism and alienation, we must examine not only the qualities of those in authority, but also those of the wider social fabric in which the citizenry is embedded. A situation where relations seem to have broken down between a government and its public— where each regards the other with suspicion and hostility—may not be remediable simply through the reform of governing institutions. It may also be necessary to cultivate the associational infrastructure of society in order to foster a public which is capable of being governed well. Without a dense network of horizontal relatedness, trust in authority—and indeed an authority worthy of trust—may have no soil in which to take root and grow.

Associations and Human Flourishing

I have used two kinds of argument in defence of the idea that the fostering of voluntarily associational activity in society should play a large part in the pursuit of sustainable development. The first was based on the idea that voluntary associations which are active in relevant areas can be extremely effective in the pursuit of sustainable development. The second argued that

156

the very existence of a dense array of associations of almost whatever type will help create a cultural climate within which the delivery of sustainable development objectives will be so much more possible, through the stimulation of trust and public-spiritedness. Whereas the first kind of argument left open the possibility that voluntary associations are of strictly delimited relevance, since they cannot totally replace the need for more large-scale and formal mechanisms of governance, the second kind brought even the latter directly into the frame, by suggesting that the civic virtues encouraged by a dense array of associational activity constitute a cultural precondition for effective governance. However, there is even a third kind of argument that might be offered in support of the role of voluntary associations in the sustainable society: one that makes the point that the very act of participating in associational activity can itself generate the kind of human flourishing which any definition of sustainable development should include.

The distinctiveness of this kind of argument can be brought out by considering two different accounts of 'happiness', which, after all, should in some sense be the end of any politics. The first account is the moral equivalent of the end-of-pipe solution much decried by environmentalist. According to this, happiness is something that lies, with luck, at the end of a process, an instrumental goal, which may be achievable by any one of a number of means. The first two kinds of argument above use such a causal, end-of-pipe view of happiness, in that they argue that voluntary associations are a means to an end of human (and non-human) happiness and flourishing, that they can deliver social and environmental goods necessary to such flourishing, and also help deliver the broader cultural preconditions of such flourishing.

But Aristotle would have us see happiness not so much as an emotion or as an end-state but as a quality which our lives may or may not possess. Happiness, according to this view, consists in the absorbed, proficient performance of an activity which is being done, not as a means to an end, but as an end in itself. Voluntary associations also produce happiness in this Aristotelian sense, in that, through engaging in shared activities with others, people come to be oriented towards the rich variety of human goods associated with that activity. Whether in a political campaigning group, an amateur football team, or a health self-help group, they find those particular kinds of pleasure and self-development which accompany activity done in some sense for its own sake. Any crude instrumentality in the initial motivation of participants so often becomes surrounded, like a piece of grit in an oyster, by a pearl of complex human engagement and loyalty. Then even the term 'association', with its connotation of merely instrumental and provisional motivations amongst participants, quickly becomes inadequate.

The satisfactions and rewards which participating in the associational life of society can bring should not just be regarded as 'useful', in the sense that they produce loyalty and enthusiasm, and thus contribute to the longevity and vigour of voluntary associations. Neither should they only be valued for their capacity to displace more consumerist forms of motivation, and thus to create the possibility of a society where satisfaction is found not through the unsustainable process of producing and purchasing more and more things, but by the pursuit of human excellence and sociality. They should also be valued in their own right, as forms of human flourishing; or, in language more familiar to sustainable development debates, as contributions to the 'quality of life'. We must not be so dazzled by the profusion of *means* at our disposal in modern society that we lose sight of the *ends* which they are supposed to be there to further. Voluntary associations can help deliver the sustainable society; they can help create its cultural preconditions; but they are also already part of it.

Biographical Note

Bronislaw Szerszynski is Research Fellow at the Centre for the Study of Environmental Change (CSEC), Lancaster University. He co-edited *Risk, Environment and Modernity* (Sage, 1996), has researched extensively into the ethics of the environmental movement, and is currently engaged in research into sustainable development, voluntary associations and global citizenship.

Notes

1 I would like to acknowledge the support of the Economic and Social Research Council, whose funding of the CSEC research programme 'Science, Culture and The Environment' made the writing of this piece possible. I would also like to thank Alan Whitehead, Robin Grove-White and the participants of the two-day workshop *Politics in Place*, held at Lancaster University on January 8th and 9th 1997, for stimulating some of the lines of thought that have led to the present chapter.
2 World Commission on Environment and Development (WCED), *Our Common Future*, Oxford, Oxford University Press, 1987.
3 Department of Transport, *Managing the Trunk Road Programme*, London, HMSO, 1995.
4 Phil Macnaghten, Robin Grove-White, Michael Jacobs and Brian Wynne, *Public Perceptions and Sustainability in Lancashire*, Preston, Lancashire County Council, 1995.
5 See also Stephen Young, *Promoting Participation and Community-Based Partnerships in the Context of Local Agenda 21: A Report for Practitioners*, Manchester University, Government Department, 1996; Michael Jacobs, *The Politics of the Real World*, London, Earthscan, 1996, pp. 96–100.
6 Dick Atkinson, *The Common Sense of Community*, London, Demos, 1994; Paul Hirst, *Associative Democracy*, Cambridge, Polity, 1994; Paul Hirst, 'Democracy and Civil Society', in *Reinventing Democracy*, ed. Paul Hirst and Sunil Khilnani, Oxford, Blackwell, 1996, pp. 97–116.
7 Macnaghten *et al.*, *op. cit.*
8 Bronislaw Szerszynski, 'The Varieties of Ecological Piety', *Worldviews: Environment, Culture, Religion*, Vol. 1, No. 1, 1997, pp. 37–55.
9 Donella H. Meadows, Dennis L. Meadows, Jorgen Randers and William W.

Behrens III, *The Limits to Growth*, London, Earth Island, 1972, p. 195; WCED, *op. cit.*, p. 23.

10 Helmuth Berking, 'Solidary Individualism: The Moral Impact of Cultural Moder-nisation in Late Modernity', in Scott Lash, Bronislaw Szerszynski and Brian Wynne, eds., *Risk, Environment and Modernity*, London, Sage, 1996, pp. 189–202. See also John Vidal's only slightly tongue-in-cheek description of road protest camps as schools for citizenship: 'The Scum Also Rises', *Guardian*, 29 January 1997, p. 17.

11 Robert D. Putnam, *Making Democracy Work*, Princeton, Princeton University Press, 1993.

12 Macnaghten *et al.*, *op. cit.*; Brian Wynne, 'May the Sheep Safely Graze? A Reflexive View of the Expert-Lay Knowledge Divide', in Lash *et al.*, *op. cit.*, pp. 44–83.

Public Opinion and the Environment

ROBERT WORCESTER

Introduction

Survey research measures five things: *knowledge*, what we know; *behaviour*, what we do; and then people's views, their *opinions, attitudes and values*. I have defined these latter terms, rather too poetically perhaps for scholarly adoption, as '**opinions**: the ripples on the surface of the public's consciousness, shallow, and easily changed; **attitudes**: the currents below the surface, deeper and stronger; and **values**: the deep tides of public mood, slow to change, but powerful.'[1]

People in Britain admit they don't know very much about the environment, and think scientists don't know very much about it either, and what is known today may or may not be proved right next year. Conflicting scientific reports abound, and public understanding of science about the environment is not helped by the media's habit of 'even-handedness', pitting the spokesperson for say 'global warming', who represents the vast majority of scientific opinion, against the scientific sceptic, who speaks for a few dissidents. Further, most people don't trust what scientists have to say about the environment anyway, and fewer trust government scientists or scientists from industry, than scientists representing environmental organisations.

People do know pretty well what they do, and what they know, so some of the data in this chapter is based not just on what people think, but on their behaviour and knowledge as well. If in 1988 14 per cent of the British public could be classified by their behaviour as Environmental Activists and this rose in 1991 to 31 per cent, and then has fallen back somewhat to around 28 per cent, and if over a third of people say they consciously 'shop green', then something has certainly happened to cause this, and it is important to understand why, and the likely political and social impact of these changes in people's behaviour.

Following the work of Abraham Maslow[2] and Ronald Inglehart,[3] who teach us that there is a 'hierarchy of human needs' (Maslow) that can be extended from individuals to cultures and nations (Inglehart), it is clear that during the economic boom years of the eighties people in the developed countries, particularly the more affluent, shifted their attention from more basic economic priorities to 'higher' concerns for quality of life. At the same time, a variety of environmental disasters—oil spills, chemical and nuclear plant accidents and pollution—were widely publicised by the media, and their effects on human and ecological health forced onto the political agenda. The rise of green political parties, and the discovery of the threats of global

© The Political Quarterly Publishing Co. Ltd. 1997
Published by Blackwell Publishers, 108 Cowley Road, Oxford OX4 1JF, UK and 350 Main Street, Malden, MA 02148, USA

warming, ozone depletion, acid rain and loss of tropical rainforests, elevated the British public's gaze above local concerns with litter and noise to encompass a global consciousness about what is happening to the environment.

Environmental Attitudes

Public concern about the environment has been consistently high for a decade. Seven people in ten believe that 'pollution and environmental damage are things that affect me in my day-to-day life'. 71 per cent reject the notion that 'Too much fuss is made about the environment these days', four in ten doing so 'strongly'. True, these results show some differences among different social groups. Three quarters of the under 55s think there is not too much fuss, but only two-thirds of the over 55s. More of those in occupational classes D and E think that too much fuss is made than in classes A and B. But a majority of all subgroups agree. (see Figures 1 and 2.)

Q. To what extent do you agree or disagree: Pollution and environmental damage are things that affect me in my day-to-day life?

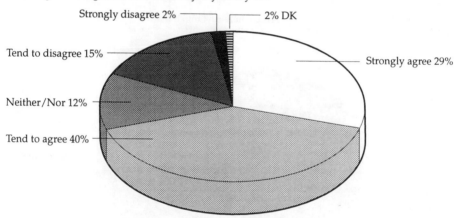

Base: 976 British adults 18+, 9-12.8.96   Source: MORI Business & The Environment Study

Figure 1

Alert politicians should take note of the increased concern about specific environmental concerns over the past decade. Work done by NOP for the Department of the Environment in 1986, 1989 and 1993 showed that concern over traffic exhaust fumes and urban smog has risen from 23 per cent 'very worried' in 1986 to 33 per cent in 1989 and 40 per cent in 1993; losing Green Belt land went from 26 per cent in 1986 to 35 per cent in 1993. On the other hand, concern over ozone layer depletion, global warming, and acid rain has

Q. To what extent do you agree or disagree: Too much fuss is made about the environment these days?

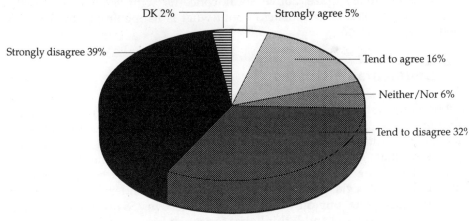

Base: 1,999 British adults 18+, 9-12.8.96   Source: MORI Business & The Environment Study

Figure 2

Q. To what extent. Do you agree or disagree: There isn't much that ordinary people can do to help protect the environment?

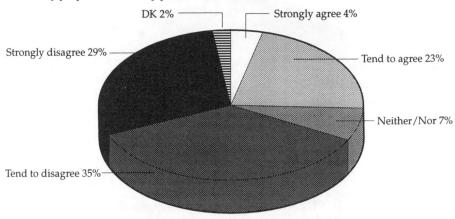

Base: 1,999 British adults 18+, 9-12.8.96   Source: MORI Business & The Environment Study

Figure 3

descreased over the period.[4] Other findings from the 1993 survey consistent with MORI's work include the finding that women tend to be more concerned than men about the environment. For 18 of the 27 selected environmental issues tested, significantly more women than men expressed concern. The greatest differences between women's and men's concern tended to be local

162

and national rather than global issues (e.g. drinking water quality, traffic exhaust fumes and urban smog, use of insecticides and fertilisers, and fouling by dogs). Yet NOP found that women were less knowledgeable than men about causes of environmental problems; for example, many fewer women (37 per cent) than men (48 per cent) identified the correct description of global warming.

MORI's survey findings show that the British public continue to reject the idea that they are powerless over environmental change. Chattering class folklore has it that the public has turned its back on the environment out of a sense of frustration that problems such as global warming and disappearing biodiversity are too remote from them, and that their own efforts, in recycling, supporting green charities, etc., are too small to make any real difference. Over the period 1993–1996, a steady two-thirds of the public (64–68 per cent) said they disagreed with the statement that 'There isn't much that ordinary people can do to help protect the environment'. Again there is an age difference: seven in ten of those under 55 disagreed, one in three strongly, but only just over half, 52 per cent, of the over 55s disagreed. (See Figure 3.)

The Environment as a Political Issue

But do these environmental concerns translate into political priorities? Over recent years the environment, pollution, conservation and allied concepts have both topped and tailed monthly measures of the public's concerns, in response to the unprompted questions ,'What would you say is the most important issue facing Britain today?/What do you see as other important issues facing Britain today?' In July 1989 'green' issues were nominated by over a third of the electorate, 35 per cent, giving it top score among all issues suggested. That month, health care/NHS was nominated by 29 per cent and unemployment by 24 per cent. Prices/inflation (17 per cent), economy/economic situation (16 per cent), and crime/law and order (15 per cent) tied for fourth place. But by November 1990 the percentage suggesting that environmental issues were among the most important facing the country had faded to under 9 per cent, and by December 1991 to just 4 per cent. The environment's rating has remained under 10 per cent since then, much less (as Figure 4 shows) than for unemployment, health care and law and order. By April 1997, the month of the general election, these three issues, along with education, were thought of as the main priorities by some ten times as many people as expressed concern over the environment.

Further, while a rise in concern about the economy, unemployment, health care and crime are all correlated positively, there is a significantly high negative correlation between nomination of these issues and mention of the environment (see Table 1). The environment is moreover correlated negatively with the Economic Optimism Index (EOI), which measures people's

Q. What would you say are the most important problems facing Britain today?

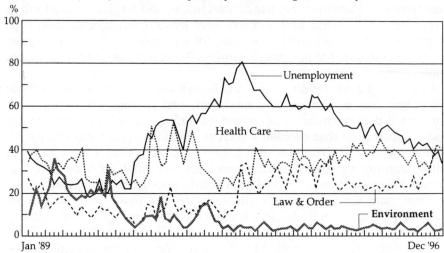

Base: c. 2,000 British public, monthly   Source: MORI/The Times

Figure 4

Table 1 Correlations: 1989–1996

Correlations	Environment	Unemployment	NHS	Crime	EOI
Environment	1.00				
Unemployment	−0.27	1.00			
NHS/Health	−0.45	0.15	1.00		
Crime	−0.53	0.51	0.31	1.00	
EOI	−0.35	0.47	0.40	0.41	1.00

Source: MORI/Times.

expectations of future economic prosperity. This confirms the claim that concern about the environment is a 'third-order' Maslovian factor, ranking after basic human needs for personal sustenance, warmth, health, financial well-being and safety, and the needs for love and esteem. People's concern for the environment rises when they feel economically more secure.

Yet these figures tell only one story. Another way of gauging public opinion towards the environment is by asking people the priority they give to the environment vis-a-vis the economy. In 1994 twice as many people (31 per cent) said they would choose to protect the environment at the expense of the economy as said they would protect the economy at the expense of the environment (15 per cent). It should be noted that this was when six people in ten were saying that unemployment was one of the most important issues facing Britain today, and 14 per cent more people believed that the general economic condition of the country would get worse over the following year than thought it would get better. By December 1996, nearly three times as many (35 per cent) favoured the environment over economic performance (12 per cent). As one would expect, this reflected a halving of concern over unemployment, the jobless figures having fallen. This 5 per cent 'swing' away from the economy and in favour of the environment is certainly in the direction one would expect from Maslow and Inglehart, and consistent with the data from other surveys.

What is perhaps more surprising is the number of people who said they believed that 'We should protect the environment at all costs, regardless of economic considerations.' These people, who might be called the 'Deep Greens', numbered some 12 per cent of the British adult population, roughly five million people. Interestingly, there are more 'Deep Greens' among older people, 14 per cent of the over 55s, compared to 10 per cent of the 18–34 year old age group. Even more surprising is that while 9 per cent of middle-class households are 'Deep Green', this attitude is held by 15 per cent of those in working class occupations. These findings suggest that while it is generally true that the environmentally concerned are younger and middle class, there is a core group of older, working class people whose values are deeply green and strongly held.

The 'Environmental Activist' and the 'Green Consumer'

For more than twenty-five years, MORI has used a 'Socio-Political Activist' typology to identify the 10 per cent or so of the public who are the 'movers and shakers' of British society.[5] In an adaptation of that concept, an 'Environmental Activist' (EA) typology was introduced in 1988, using a behavioural scale of environmental actions (see Table 2). It has proved powerfully predictive, enabling environmental behaviour in Great Britain to be tracked consistently over the past eight years, and has now been extended to over 30 countries.

Over the period 1988–96, there was a sharp rise in the proportion of the British public who claim to have done various green activities, whether this is minor activities such as walking in the countryside or watching environmental programmes on TV, or more far-reaching behaviour such as 'green purchasing'—an activity whose practitioners have doubled in number over the eight years. The proportion of Environmental Activists (defined as those who have taken five or more of the listed actions) has more than doubled, from 14 per cent of the adult population in 1988 to 31 per cent in 1991 and 28 per cent in August 1996. Thus over the last few years the proportion of the British public whose actions suggest we can call them 'green' has remained more or less constant at a quarter of the adult population, or a little over ten million people.

There are twelve activities in the Environmental Activism scale. In 1995 the

Table 2 British Green Activism

Q. 'Which, if any, of the following things have you done in the last year or two?' (%)

Date of Fieldwork	1988	1989	1990	1991	1992	1993	1994	1995	1996
Read/Watched TV about wildlife/conservation/natural resources/3rd World	79	80	84	87	85	83	81	85	85
Walked in the countryside/along the coast	72	75	77	81	75	77	79	81	80
Given money to or raised money for wildlife/conservation or 3rd World charities	28	45	48	57	49	54	52	56	51
Selected one product over another because of its environmentally friendly packaging, formulations or advertising (GCs)	**19**	**47**	**50**	**49**	**40**	**44**	**40**	**41**	**36**
Requested information from an organisation dealing with wildlife/conservation/natural resources/3rd World	7	14	15	15	12	13	12	15	16
Subscribed to a mag. concerned with wildlife/conservation/nat. resources/3rd World	8	14	13	15	10	13	11	13	14
Been a member of an environmental group/charity (even if joined more than 2 years ago)	6	8	9	13	8	12	11	12	10
Visited/written a letter to an MP/councillor about wildlife/conservation/natural resources or the 3rd World	5	5	5	5	4	5	6	6	6
Campaigned about an environmental issue	4	4	6	5	3	5	5	6	5
Written a letter for publication about wildlife/conservation/natural resources or the 3rd World	2	2	2	3	2	2	2	2	3
Used lead-free petrol in your car	12	22	25	35	37	42	45	46	51
Environmental Activists (EAs) (5+ Activities)	**14**	**20**	**25**	**31**	**23**	**28**	**26**	**29**	**29**

Base: c. 2,000 British adults
Source: MORI Business & the Environment Survey

average person reported that he or she had done 3.17 of these activities in the previous year. Women (3.24) had done a few more than men (3.08). Those in occupational class AB (managerial and professional) had done on average 3.85, those in C1 (white collar) 3.34, those in C (skilled working-class) 3.04 and the average DE (unskilled) 2.58. In the latest (1996) survey, four in ten (43 per cent) of ABs qualify by their behaviour as Environmental Activists; ABs are almost three times as likely to do five or more of the named activities as DEs.

MORI has defined 'Green Consumers' as those who have selected one product over another because of its environmentally-friendly formulation or packaging. These number more than a third (36 per cent) of British adults. Most are women (58 per cent), middle-class (59 per cent) and younger (62 per cent are under 45, against 53 per cent of the total adult population). With an adult population of some 42 million adults, each single percentage point represents some 420,000 consumers. Therefore, when 24 per cent of the general public claim to 'Avoid using the services or products of a company which you consider has a poor environmental record', this suggests that more than ten million consumers are at least potentially willing to using their purchasing power to affect the environmental behaviour of manufacturers. This could clearly be a powerful consumer force. (See Table 3.)

Who is Trusted on Environmental Issues?

In 1995 MORI found that nearly half of the British public (45 per cent) agreed with the statement 'I don't fully understand environmental issues'. Moreover, more than a third, 36 per cent, agreed in 1996 that 'Even the scientists don't really know what they are talking about when it comes to the environment'. This included a third of the most educated group, the ABs. It was before both the Brent Spar and BSE episodes.

As Figure 5 shows, the public's confidence in the information they receive on environmental issues is governed to a considerable extent by its source. Only a third of people trust what scientists working for the government have to say about the environment, with industries' scientists commanding the trust of less than half of the British public. By contrast, three quarters of the population have a fair amount or great confidence in the pronouncements of scientists from environmental organisations.

But these figures are not static. After a period of little change between 1993 and 1995, 1996 saw a sharp drop in confidence (by 6 per cent) in government scientists, possibly the result of the BSE affair. Similarly, perhaps reflecting the arguments over Brent Spar, 1996 saw a fall in confidence (of 3 per cent) in the information put out by industries' scientists. Interestingly, whereas the 'swing' against industry scientists has been most marked among DEs (12 per cent) and readers of the popular press (5 per cent), by contrast, readers of the quality press have not swung at all, and ABs have swung 1.5 per cent towards believing scientists working in industry. It is, however, the environmental groups which have

Table 3 British Green Consumers

Q. 'Which, if any, of these things do you do or have you done in the last 12 months as a result of concern for the environment?'

	ALL 1990 %	ALL 1991 %	ALL 1992 %	ALL 1993 %	ALL 1994 %	ALL 1995 %	ALL 1996 %	GC† 1996 36%
Buy 'ozone friendly' aerosols/avoid	73	71	65	71	66	70*	70*	80*
Buy products made from recycled material	40	52	51	54	52	53	50	75
Buy household, domestic, or toiletry products that have not been tested on animals	43	51	47	51	52	53	49	73
Buy products which come in recycled packaging	41	55	52	50	45	49	46	71
Regularly use a bottle bank	39	39	43	46	44	52	48	59
Buy free-range eggs or chickens	44	46	44	45	49	49	51	67
Keep down the amount of electricity and fuel your household uses	44	44	42	51	53	52	47	59
Buy products which come in biodegradable packaging	26	34	29	36	33	36	34	57
Send your own waste paper to be recycled**	31	36	36	36	38	42	50	63
Avoid using chemical fertilisers or pesticides in your garden	41	38	31	38	40	37	43	57
Buy 'environmentally' phosphate-free detergents or household cleaners	38	37	35	40	37	39	34	54
Buy food products which are organically grown	25	28	24	27	26	28	28	46

Avoid using the services or products of a company which you consider has a poor environmental record	23	19	16	22	22	23	24	46
Keep down the amount you use your car	19	19	13	16	17	22	20	29
Have a catalytic converter fitted to your car	9	7	6	9	9	11	14	19
Avoid buying chlorine bleached nappies	13	10	7	7	6	5	6	10
TOTAL	549	586	541	599	589	621	614	865
AVERAGE	36.60	39.07	36.07	39.93	39.27	41.40	38.38	54.04
Take your own waste paper to be re-cycled or have it collected for recycling**								
Regularly use a can bank					30	34	36	47
Have put in loft insulation/extra loft insulation					18	29	31	35
Buy low energy light bulbs for your home					16	24	24	32
Buy fridge/freezer with reduced CFCs					n/a	20	23	31
Joined an environmentally/ethically friendly bank or taken out an environmentally/ethically sound investment						3	4	7
None					6	8	6	6
Don't know/can't remember					1	1	*	*
Green Consumer Activists (8+)				42	39	42	41	69

Base: c. 2,000 British adults (2,048 in 1995).
*estimated.
**question reworded in 1995/1996.
Source: MORI Business & the Environment Survey.
†GC = Green Consumers.

Q. How much confidence would you have in what each of the following have
to say about environmental issues?–A great deal/A fair amount/Not very
much/None at all/Don't know (omitted).

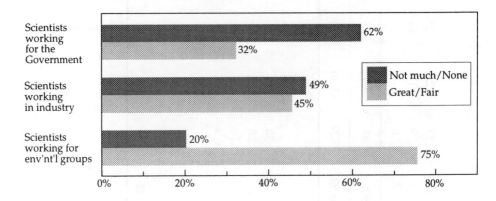

Base: 1,999 British adults 18+, 9-12.8.96   Source: MORI Business & The Environment Study

Figure 5

suffered the biggest decline in confidence in this period, a fall of seven
points, from 82 per cent to 75 per cent. There have been no winners, it
seems, from the scientific controversies of recent years.

When it comes to taking measures to protect the environment, up to 1995
more people were on the side of Brussels than Whitehall. When asked in 1995
'Who would you trust more to make the right decisions about the environment?', 39 per cent said they would trust the European Union, 32 per cent the
British Government. However this reversed in 1996, with 38 per cent trusting
the Government more and 33 per cent the EU. (The proportion saying they
would trust neither remained the same at 20–21 per cent.) This change reflects
a general strengthening of Euro-scepticism over the year, although it may also
owe something to the EU's handling of the BSE crisis. There are marked
differences between different social groups, however. Older people, those 55
and over, back the British Government by more than two to one; young
people (those between 18 and 24) are ten points more on Europe's side. Those
in white collar and skilled manual jobs are marginally more confident in the
EU, while the DEs favour the British Government by 18 per cent. (See
Figure 6.)

The Party Political Implications

What are the implications of all this for electoral politics? Does the environment fade away at election time, as it has in the past, swept aside by the four
horsemen of unemployment, health care, education and law and order? Or

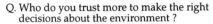

Q. Who do you trust more to make the right
decisions about the environment ?

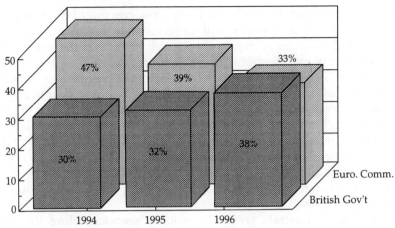

Base: 1,999 British adults 18+, 9-12.8.96   Source: MORI Business & The Environment Study

Figure 6

does it become a cutting edge issue, important enough for some to turn their
vote?

There are four criteria that give an issue 'bite'. First, it must be salient to
the elector, an issue of sufficient concern to help determine party choice. On
this basis the environment is a salient issue for about one in five of the
electorate. In the two years before the 1997 general election, when asked
'Looking ahead to the next General Election, which, if any, of these issues
do you think will be very important to you in helping you to decide which
party to vote for?', a consistent one in five, 20 per cent, said that 'the
environment' would be important to them in casting their vote. This
represents around ten million people. Slightly more of these people were
women (57 per cent) than in the electorate as a whole. They also tended to
be younger (37 per cent as against 33 per cent of the electorate) and more
middle class (58 per cent of ABC1s, as against 48 per cent). Thus while the
public do not think of the environment as necessarily among the most
important issues facing the country today, they do think that it is important
to them personally in an electoral context.

Second, the issue must be one on which there appear to be discernible
differences between the parties. MORI asked those one in five for whom the
environment was salient which party had the best policies on 'protecting the
natural environment'. Nearly one third (31 per cent) felt that no party had the
best policies, or they didn't know which party had the best. Though salient,
environmental issues were clearly unlikely to influence these people's votes
unless the parties focused on the issue. Among those who did express a
preference for one party's policies, 10 per cent favoured the Conservatives,

20 per cent Labour, 19 per cent the Liberal Democrats and 20 per cent the Green Party.

The high number favouring the Green Party's policies might be thought to give them electoral encouragement. But unfortunately for them, the third criterion which makes an issue electorally significant is that the voter must believe that the party whose policies are favoured would and could do something about it. The Green Party's weak electoral position means that it is widely regarded as failing this test. So although two million people appear to support its environmental policies, it is ruled out as a beneficiary of their votes. In fact, throughout 1996, the Green Party registered a voting intention of just one person in a hundred. These supporters were not typical of the electorate as a whole. 62 per cent of them were under 35 (compared with a third of the electorate as a whole), 61 per cent middle class (compared with 48 per cent of the electorate) and over half, 55 per cent, lived in the South of England, compared with four in ten electors.

Since the Liberal Democrats were also widely seen as failing the third criterion—being highly unlikely to form a government—this left Labour and the Conservatives apparently competing for the votes of those for whom the environment was an electorally significant issue. Labour's two-to-one policy lead over the Tories suggested that they would be the principal beneficiaries, especially as the fourth critera of electorial 'bite' is confidence that the party in power *would* do something about the environment and after 18 years the public's patience with the Tories had run out. In these circumstances the low profile given to the environment by Labour in the 1997 election campaign was perhaps surprising. With ten million potential voters saying that they would put the environment among the top two or three issues determining their vote, and about one million indicating effectively that it could make them vote Labour, the environment was perhaps not as electorally irrelevant as Labour's campaign strategists appeared to think. Given the continuing concern about environmental issues shown by the British people, Labour would be similarly unwise to downplay its importance in government, and to give it credit, it hasn't—so far.

Biographical Note

Robert M. Worcester is Chairman of MORI and Visiting Professor of Government at the London School of Economics and Political Science. He is a Vice President of the Royal Society for Nature Conservation/Wildlife Trusts.

Notes

1 Robert M. Worcester, 'Reflections on Public Opinion and Public Policy', ESOMAR/ WAPOR Congress, Copenhagen, Denmark, 18 September 1993.
2 Abraham Maslow, 'Theory of Human Motivation', *Psychological Review*, 50, 1943, pp. 370–96.

3 Ronald Inglehart, *Cultural Shift in Advanced Industrial Society*, Princeton, N.J., Princeton University Press, 1990.
4 Department of the Environment, *Statistical Bulletin*, 1994, Appendix A, pp. 1–46.

Green Politics and Parties in Germany

DETLEF JAHN;

Green parties are a product of new political concerns, but even more of national political contexts.[1] These factors then influence each other. This is particularly true for the German Green Party *Bündnis 90/Die Grünen* ('Alliance 90/The Greens', referred to here simply as the Greens.) To understand the present situation of Green politics and the Green Party, it is important to recall the special conditions prevailing in Germany, and in particular the extent to which the political environment has changed since German unification.

Circumstances and Conditions for Green Politics in Germany

The key to the German Greens lies in their ideological background, the product of the post-war German situation.[2] The presence of Communist Eastern Germany, coupled with an authoritarian state culture (*Obrigkeitsstaat*), resulted in West Germany in a policy hostile to everything associated with left-wing ideology. This led to the abolition of the Communist Party (KPD) in 1956. More important, it led to the radical decree (*Berufsverbot* or *Radikalenerlaß*) introduced by Chancellor Willy Brandt in the early 1970s, which forbade people of radical left and right parties to be employed in the public sector. The decree was most frequently exercised on the left. It caused many left-wing Germans, frustrated by this first social democratic govern-ment, to feel they had no political home. At the same time, the economic prosperity of the late 1960s and early 1970s proved a breeding ground for young people who rejected the materialism of German society. Protests against nuclear power stations and other large scale developments were met with violence on the part of the state, confirming its authoritarian stance. It was in this context that the Greens emerged, integrating the ideologically homeless on the left with the anti-state sentiments and anti-establishment initiatives of the young. The Party became the voice of social minorities, advocates of equal rights for women, and champions of egalitarian principles and political morality.

Thus the Greens came to represent a small but significant part of the German population. But at the same time these circumstances led inevitably to internal tension. This crystallised into the fight between purist *Fundis* (fundamentalists) and pragmatic *Realos* (realists), which split the party into two competing factions from the early 1980s. To describe this complicated ideological split in simple terms one can say that the *Realos* favoured parliamentary politics and co-operation, including formal coalitions, with

 Published by Blackwell Publishers, 108 Cowley Road, Oxford OX4 1JF, UK and 350 Main Street, Malden, MA 02148, USA

the SPD. In sharp contrast, the *Fundis* stressed extra-parliamentary action such as demonstrations and civil disobedience, and were only prepared to co-operate with the SPD on a strictly issue-by-issue basis. Each camp developed its own intra-party organisation and had its own party leadership.

The German Greens are therefore a very special creation. Ecology is only one factor in their identity: left-libertarian attitudes are at least as important. A number of structural factors have facilitated their success. First and probably most important has been the electoral system. In Germany, unlike the UK where a party must receive the majority of votes in a constituency, there is a combination of majority and proportional systems, and the proportional aspect dominates. Every party receiving more than five per cent of national support gets parliamentary seats in proportion to its vote. This five per cent threshold has been very important. It was high enough to force unity on the various Green factions; without it, Green groups and parties would have been fragmented. But it was low enough to allow representation in Parliament at an early stage.

The second factor is the refund of electoral campaign costs. These are distributed to parties which receive more than 0.5 percent of the votes. This helps to explain how an initially small party like the Greens, lacking financially strong members or associates, managed to build up a network of party offices and electoral campaign organisations. Again, this is in sharp contrast to the situation in the UK, where elections are financially ruinous to small parties. Finally, the federal structure of the German political system provides favourable opportunities for new parties to develop. This last factor was of particular importance in the early 1990s, when the Greens were only represented by nine MPs (from the East German Alliance 90) in the *Bundestag*, but still had significant representation in the parliaments of individual *Länder* (states).

Electorally, the Greens have been increasingly successful. They first participated in federal elections in 1980, when they won 1.5 per cent of the vote and no seats. Their first parliamentary success came in in 1983, when they won 5.6 per cent and 27 seats, and this rose to 8.3 per cent and 42 seats in 1987. German unification proved a setback and the West German Greens failed to overcome the five per cent threshold in the 1990 election. But in 1994 the Greens were back again on the national stage with 7.3 per cent of the vote and 49 seats. The same trend has occurred in elections to the European Parliament. The last European election in 1994 saw the Greens gain their highest ever share of the national vote, 10.1 per cent.

Since the early 1980s the Greens have also been successful in state elections. Currently they are represented in all the state parliaments of the former West German states and in Saxony Anhalt in the East (for the difference between West and East Germany see below). Indeed at the end of 1996 they were in governing coalitions with the SPD in Saxony Anhalt, North-Rhine-Westphalia, Hesse and Schleswig-Holstein. Least noticed of all, but in many ways the cornerstone of Green strength, has been their performance at local level. With more than 7,000 representatives in cities, towns and counties, the Greens have created an organisational infrastructure which has helped them to survive even in harsh times. Participating in a variety of formal and informal council coalitions, they have achieved most in areas such as recycling programmes, controls on traffic, women's refuges and multi-cultural festivals.

The Impact of the Greens

The Greens often talk of the established parties in terms of *themenklau*, or 'stealing issues'. Although the Greens have generated no major national legislation of their own, their impact on German political culture has been substantial. Particularly in the fields of gender equality and environmental policy, they have clearly influenced the policies of the established parties. In the mid-1990s, for instance, even the Christian Democrats introduced a quota for women on public boards, which the Greens had introduced in their own party organisation two decades before. The Greens have been instrumental in keeping the issue of the environment on the political agenda. They have left their special mark in preventing new high-speed highways and nuclear power plants and in the redesign of waste disposal plans. The Greens have had particular influence over policy in coalition governments at state level. They have usually controlled the ministries of women's affairs and the environment. But although the Greens focus on environmental and gender questions they are not a single issue party. They have developed strong expertise in social policy and have recently also increased their economic profile.

The rising Green vote has significantly affected the established parties. The Greens have displaced the liberal Free Democrats (FDP) as the third strongest party. They have reduced the power of the Social Democrats (SPD), forcing them in several states to create coalitions with Green participation. To try to counter their support, each of the other parties has veered between opposing Green positions and adopting them. This has inevitably led to virulent internal argument within these parties—a situation which is in any case more rule than exception in Germany, at least for the SPD and the FDP.

The internal conflict within the SPD, in contrast to their counterparts in most other European countries, is institutionalised in various powerful and well-resourced intra-party organisations. In the context of Green politics, the split in the (West) German Left between 'traditional' and 'post-material' factions is particularly important. These two Lefts have some common concerns, such as social justice and equality, but the disagreements between them cause severe conflicts when it comes to questions of economic growth and environmental protection. This division of the Left cuts across the SPD, and over some issues effectively paralyses the party.

At the federal level, the attitude of the SPD towards the Greens has shifted significantly since the early 1980s. Helmut Schmidt, SPD Leader and Chancellor until 1982, made economic growth his clear priority, with only minor concessions to ecological concern. He was openly against the peace, anti-nuclear and environmental movements and the Green Party, which more or less represented them. Indeed his hostility was an important source of the Greens' early growth. Schmidt's successors, Helmut Rau and Hans-Jochen Vogel, were also committed to the SPD's traditional values and to the outlook of its core support among industrial workers. All this changed with the appointment of Oskar Lafontaine as prospective Chancellor in the 1990 election. Lafontaine confronted head-on the challenge of restructuring industrial society in terms of ecological concerns. But he was heavily criticised within his own party for being too green and the SPD's defeat in the first all-German election was caused in part by this internal dissent. Lafontaine's successors have gradually given up prioritising environmental concerns in the party's platform. Björn Engholm took the first step in this direction and Rudolf Scharping stood substantially closer to the 'materialist' side than to the Green one.[3] Lafontaine's recent comeback may indicate that co-operation between the SPD and the Greens is once again more likely; but it also means that competition between the Greens and the SPD for the 'green vote' will inevitably increase.

The emergence of the Greens has also had a profound impact on the Free Democrats. Formed after the Second World War to unify social-liberal and national-liberal streams, the FDP shifted its allegiance in 1982 from the SPD, whom it supported in coalition in the 1970s, to the Christian Democrats (CDU/CSU), enabling Chancellor Kohl to form his long-standing coalition government. But the rising star of the Greens has caused the FDP's own position to wane. This is because the Greens ' . . . attract electors who had previously formed the key "target groups" for the Free Democrats. Since about 1978, the two parties have been struggling against one another to win over young, well-educated voters who mostly live in urban areas and whose partisan ties are less deeply embedded and more fickle than older, less educated people.'[4] In the 1980s this competitive relationship between the FDP and the Greens resulted in a certain progressiveness in the FDP's environmental policy. But since unification the FDP has made a profound change of direction, increasingly looking for a political niche to the right of the CDU/CSU. This ideological shift is mainly inspired by a Thatcherite form of economic liberalism. However, it has also led to a revival of a national-liberal ideology which was prominent before (and was meant to have been eradicated after) the Second World War. Both these streams oppose the environmental and left-libertarian ideology of the Greens. The shift in the character of the FDP has changed its relationship with the Greens from co-operation to sharp conflict.

Finally, even the Christian Democrats have not been immune to the ecological challenge. Led by its social-religious wing (which is lacking in 'purer' conservative parties such as the British Tories), the CDU promoted socially and environmentally liberal politics throughout the 1980s. However since unification this group has been opposed by the low-tax and law-and-order faction which has gained power in the party. The party's environmental policy has been weakened as a result. Meanwhile the only party that seems to have been closed to Green concerns altogether is the CDU's sister party in Bavaria, the CSU.

Re-Orientation of the Greens After Unification

German unification changed the situation of the German Greens fundamentally. Again, as in its formative years, the political context had a profound impact, not only on the political opportunities available to the party, but on its very structure.

First of all there is no doubt that German unification resulted in much less attention being paid in German politics to green issues. This trend went hand in hand with an increase in the salience of the economy and 'national identity'. The trade-off between economy and environment became more acute after unification because of the relative poverty of the former East Germany, and the government of the CDU/CSU and FDP inevitably gave economic growth top priority. The sharper conflict between economic and environmental concerns also increased the tension between the factions within the SPD, strengthening the position of the industrial/labour wing. Co-operation between the SPD and the Greens in *Länder* governments has consequently declined, most dramatically in their coalition in Nordrhine-Westfalia.

This clash between the SPD and Greens is the more significant because the Greens have moderated their policy stance considerably during the last few years. There were two internal reasons for this. First, the struggle between Fundis and Realos was finally settled—in favour of the Realos. This occurred mainly because of the increasing opportunity for the Greens to form coalition governments with the SPD at Länder level in the late 1980s. Second, the Greens' failure in the federal election in 1990 came as a considerable shock, and led to fundamental organisational changes. The compulsory rotation of leaders was abolished, the 'incompatibility rule', which prohibited elected representatives from holding party office, was toned down (except for members of the Bundestag), a procedure for postal ballots was introduced, and the appointment of posts on the national steering committee was assigned to the regional executives.

The moderation of the Green ideology was reinforced by the merger with the East German 'Alliance 90' in May 1993.[5] The 'citizen initiatives' of Alliance 90 emerged for quite different reasons from those of the 'post-materialist' Greens in West Germany. Their protest was against the East German state and the Communist Party (the Socialist Unity Party of Germany or SED) which ran it. Reacting to the closed political system of the former GDR, Alliance 90 stressed the importance of 'dialogical politics'—of open debate in civil society. They were particularly concerned to reject identification as a left-wing party. In some states (e.g. Saxony) Alliance 90 was even willing to enter into coalition with the Conservative Party (CDU) and to co-operate with the authoritarian Ecological Democratic Party (ÖDP).[6] During the negotiation process with the West German Greens, it was Alliance 90 which demanded an acceptance of the political system of the Federal Republic of Germany and of the free market economy.

Not everyone in the West German Greens is happy about this, however, and though the conflict between Fundis and Realos has diminished, there is still disagreement about the Greens' political strategy. There are particular conflicts between the parliamentary group (*Bundestagsfraktionen*) and the party organisation, a well known phenomenon of German party politics. Since the Greens were returned to the Bundestag the parliamentary group, led by Joschka Fischer, has been the more powerful body. His pragmatic stance is often at odds with the more left-wing line taken by the party organisation.

West and East Germany: Two Different Worlds for the Greens

The German Greens have always done best in elections in economically prosperous states with a high degree of urbanisation, a high number of students, and where the Party is well organised and has a chance of changing the power relations in parliament. However, since unification most of these factors have changed for the worse.[7] As economic recession has brought the 'materialistic' issues of unemployment and economic growth to the fore, the Greens' poll ratings have declined even in the wealthy states. More important, however, are the structural changes resulting from German unification. East Germany is not only considerably poorer but also much more agricultural and rural than the West, and its voters have different concerns and different ideological outlooks.

The second all-German election of 1994 would appear to indicate that the Greens have a stable place in German politics. However, a closer look qualifies this impression. In Germany as a whole, the Greens received 7.3 per cent of the vote. This is less than the 8.3 per cent they won in the last election before unification in 1987. But the significant difference is between the results in the West and the East of the country. In the West the Greens obtained 7.9 per cent, coming close to their support in 1987. In contrast, in the East their support decreased by almost 2 percentage points to 4.3 per cent. Both the SPD and the PDS (the Party of Democratic Socialism, the reformed Communist Party) were more successful in gaining votes from the East German population, increasing their shares by 7.2 and 8.7 per cent respectively, to 31.5 and 19.8 per cent.

The different outcomes in East and West Germany are even clearer at state level. The Greens lost all their seats in the East German *Länder* parliaments except one, Saxony-Anhalt (where the Greens are the junior partner of a minority coalition government with the SPD, supported by the oppositional

PDS). In these states the PDS obtained an average of almost 20 per cent, increasing their share of the vote by 6.7 percentage points. In sharp contrast, in the West German states, the Greens improved their position. They now have representatives in all West German *Länder* parliaments.

So the Greens are at present a party of West Germany, like the FDP. In recent opinion polls (December 1996) their rating in all Germany is about 12 per cent, but again this hides much higher support in the West. Their poll strength is partly due to the charismatic Joschka Fischer—who is widely perceived as the opposition leader in the *Bundestag*—and partly reflects the poor current performance of the SPD. The question remains, however, how the Greens can incorporate the voters of the East without losing, or frustrating, their core support in the West.

Green Potential and Future Prospects

The average Green voter is a product of the 'protest generation' of the late 1960s and 1970s, born in the 1950s or 60s, highly educated, urban, and a white-collar worker.[8] Green voters naturally emphasise their concern about environmental issues. But they also typically hold other distinctive positions. They consider the 'asylum question' and immigration policy (on which they have liberal, anti-racist views) more important than the population in general, and they are more concerned about the re-emergence of German nationalism. Both issues are particularly important in Germany because they carry echoes from the country's Nazi past. Green core-voters are more positive about European integration than the population as a whole. They have less trust in politicians and tend to be highly critical of the prevailing political system.

But such people do not make up a major proportion of the German electorate. The Greens' core-voters, those loyal to the party, are estimated at around three per cent. Mobilising potential sympathisers, who hold substantially the same views, brings the Greens to around eight per cent. But beyond this it is not easy to predict the potential for Green votes. Some commentators estimate that the Greens have the potential to reach 20 per cent.[9] But this will depend as much on the strategic moves of their closest party competitors as on their own positioning.

The problem is that, while the three parties of the left (the PDS, the SPD and the Greens) vary in attraction over time, their relation seems to be basically a zero-sum game. That is, one party's gain is made at the expense of the others rather than of the CDU or FDP. For the two smaller parties, the Greens and the PDS, this battle becomes effectively a question of survival. It means that the future of the Greens can only be evaluated by looking at different scenarios for the relations between, and developments in, the SPD and PDS.

One scenario is that the SPD takes a more radical turn, expressing greater concern for the under-privileged and the environment. If this happened it is likely that the SPD would gain votes from the PDS and the Greens, risking the possibility that one or both of these parties would fall short of the five per cent threshold. Proportionally, this would increase the voting share of the CDU/ CSU and the FDP. An absolute majority for the SPD is very unlikely, but a more radical line would make the SPD poorly placed for a coalition with the CDU. The left as a whole would not gain. Another scenario is that the Greens moderate their own politics further, making them more attractive as a coalition partner to the SPD. However, this might alienate their core anti-establishment support, leading to a strengthening of the position of the PDS. It might leave the SPD and Greens in partnership, but in a minority, and the CDU/CSU and FDP again therefore in government. A third scenario would see the Greens becoming more radical. This might pull support in from former PDS voters, but it would make the Greens unsuitable for coalition with the SPD, again leaving the left in opposition.

To overcome the zero-sum problem the Greens need to attract votes from parties other than the SPD and PDS. One possibility would be the collapse of the FDP, with its voters defecting to the Greens and SPD. A short while ago this did not look implausible, but the FDP has regained strength in the latest state elections and it has now become much less likely. We are left then with the possibility that the proportion of core Green voters could itself increase, making Green support less dependent on defections from other parties. But unfortunately for the Greens, this does not seem likely. The environment does not look like becoming a major issue again in the near future and the special conditions which created the original core support for the German Greens (student protest, economic prosperity, repressive state responses) show little sign of returning.

Of course, political reality is often steered by chance, by sudden options and constraints not previously foreseen. Though highly unlikely at the moment for historical reasons, it is not impossible that co-operation between the Greens and PDS could emerge. Indeed this would be the obvious way to overcome the complementary regional weakness of the two parties (the Greens in the East, the PDS in the West). However, the most likely alternative to the current government remains a coalition between the SPD and the Greens. For supporters of such a prospect, it has to be said that the main concern is the ability of the SPD to govern, not the condition of their potential partner.

Biographical Note

Detlef Jahn is Professor of Political Science at Nottingham Trent University, where he is carrying out research on comparative environmental politics. He also works on election campaigns and European Union referendums. He is the author of *New Politics in Trade Unions* (1993).

Notes

1 A comparative analysis along these lines has been conducted in Detlef Jahn, 'The Rise and Decline of New Politics and the Greens in Sweden and Germany: Resource

Dependence and New Social Cleavages', *European Journal of Political Research*, 1993, pp. 177–94.
2 Especially thoughtful accounts of the ideological roots and streams of the German Green Party can be found in Andrei S. Markovits and Philip S. Gorski, *The German Left: Red, Green and Beyond*, Oxford, Polity Press, 1993; Helmut Wiesenthal, *Realism in Green Politics: Social Movements and Ecological Reform in Germany*, Manchester, Manchester University Press, 1993; and E. Gene Frankland and Donald Schoonmaker, *Between Protest and Power: The Green Party in Germany*, Boulder, Westview, 1992, particularly chapter 2.
3 One indicator of Scharping's position is the coalition choice under his leadership in Rhineland Palatinate in April 1991, where a very moderate Green Party received 6.4 per cent of the vote. In this case, the SPD opted for the FDP as junior partner although the latter had won fewer votes than the Greens.
4 Emil J. Kirchner and David Broughton, 'The FDP in the Federal Republic of Germany: The Requirement of Survival and Success', in Emil J. Kirchner, ed., *Liberal Parties in Europe*, Cambridge, Cambridge University Press, 1988, p. 75.
5 An analysis of this process can be found in: Detlef Jahn, 'Unifying the Greens in Unified Germany', *Environmental Politics*, 1994, pp. 312–18.
6 Christoph Hohlfeld, 'Die Grünen in Ostdeutschland', in Joachim Raschke, ed., *Die Grünen: Wie Sie Wurden, Was Sie Sind*, Köln, Bund-Verlag, pp. 395–416.
7 This is based on a statistical analysis at the *Länder* level. For further details see Detlef Jahn, 'Changing Political Opportunities in Germany: The Prospects of the German Greens in a United Germany', *Current Politics and Economics of Europe*, 1994, pp. 209–26.
8 Rüdiger Schmitt-Beck, *Vor dem Wahljahr 1994. Wählerpotentiale von Bündnis90/Die Grünen in West- und Ostdeutschland*, manuscript, September 1993.
9 Joachim Raschke, *Die Grünen: Wie Sie Wurden, Was Sie Sind, op cit.*

Greening and Un-Greening
The Netherlands

PAUL LUCARDIE

A SHORT visit to The Netherlands may create the impression that it is a green country; not only because of its green meadows and pastures, but also because of the environmental concerns expressed by the people and the government. One can often observe citizens bringing used bottles to bottle banks, and many urban citizens seem to prefer the bicycle to the car. Moreover, the government impressed citizens as well as foreign experts with its National Environmental Policy Plans.

Yet first impressions are not always reliable. Close analysis suggests that environmental consciousness in The Netherlands is superficial and subject to considerable fluctuations. To use a 'green' image: it blossomed rapidly in the early seventies, carried some fruit in the late eighties and withered away like autumn leaves in the nineties. Though environmental agencies have their own momentum—like other government agencies—they cannot do without public support. As the Dutch political system is quite open and accessible to new ideas and new parties, public opinion is translated quickly into government policy. Thus the early 'greening' of Dutch politics was followed by some 'un-greening' in recent years. Both processes will be described and explained here.

Environmental Politics: A Short History

The Netherlands is the most densely populated country in Europe, fifteen million inhabitants living on less than 15,000 square miles of land: more than 1000 on a square mile. Most of them live in the urbanised western provinces, leaving still almost two-thirds of the country's surface to agriculture. Only 12 per cent of the land has not been cultivated or built up in some way; but even those 'natural' forests, marshes, sand dunes and wetlands are planned and controlled by man. Like historic and cultural monuments, 'natural monuments' are protected against wanton destruction or exploitation. For this purpose a Society for the Protection of Natural Monuments was founded as early as 1905, at a time when urbanisation and industrialisation began to threaten the remaining Dutch lakes, woodlands and marshes. The Society still exists today with 725,000 members in 1995.

However, until the late sixties the Dutch people did not perceive their fragile 'natural' environment as a political problem. The first group to politicise environmental issues was Provo: a small but influential movement

© The Political Quarterly Publishing Co. Ltd. 1997
Published by Blackwell Publishers, 108 Cowley Road, Oxford OX4 1JF, UK and 350 Main Street, Malden, MA 02148, USA

of neo-anarchist youth which held street 'happenings' in Amsterdam and a few other cities between 1965 and 1967. Provo criticised the air pollution caused by cars as well as oil refineries and other industries. Moreover, it developed 'white plans', simple if not simplistic solutions to these problems: 'white-bicycles' provided by municipal services should replace cars in downtown areas; polluting industries should pay additional taxes for the right to use 'white chimneys'. Provo's concerns, though not their solutions, were taken up soon by some political parties, mainly on the Left: the Pacifist Socialist Party, the Radical Party and Democrats 66. They were still fairly small groups, however, and concerned about many other issues apart from the environment.

In the early seventies, the publication of *Limits to Growth*, the report to the Club of Rome, hit Dutch public opinion like a bomb-shell. More copies were sold in The Netherlands than in all other countries put together. Environmental problems were given first priority by 22 per cent of voters in the early 1970s—more than any other problem except for housing shortage—whereas in 1967 only one per cent had considered environmental questions this important.[1] The government established a Department of Public Health and Environmental Protection in 1971.

Later in the seventies, an anti-nuclear movement emerged in The Netherlands, as elsewhere. Rather than confront the movement, the Dutch government tried to pacify and accommodate it. Between 1981 and 1984 the government organised a 'society-wide discussion' (*Brede Maatschappelijke Discussie*) about its plans for new nuclear power stations. The accident at Chernobyl proved a powerful 'post-hoc' argument in the 'society-wide discussion': all plans for new reactors were shelved. In 1994 the Dutch parliament finally decided to close the remaining two power plants in the near future.

Yet apart from the question of nuclear energy, environmental issues did not play a significant role in Dutch politics during the late seventies and early eighties. One might say it had become a latent concern of most citizens: when questioned the Dutch would still express serious worries about the environment. For example, in 1985 only 8 per cent agreed with the statement 'If protection of nature has to cost me money, I am not interested'.[2] Yet in daily life as in political life other problems like housing, health, work or unemployment seemed more important.

Around 1989 this changed again—in The Netherlands as in other countries—only more so. World-wide discussion of global warming (the greenhouse effect) and the 'hole' in the ozone layer contributed to this renewed environmental consciousness. A report of the National Institute for Public Health and Environmental Protection, entitled *Zorgen voor Morgen* ('Worries for Tomorrow') hammered it home in 1988. The government adopted a National Environmental Policy Plan in 1989, which addressed—rather ambitiously—all environmental problems in an integrated approach. Sustainable development was the central aim, to be achieved by 2010. Specific targets were set for each sector or target-group: agriculture, transport and traffic, energy, industry and refineries, construction, research and education, consumers and retail trade. If the sectors were well-organised, the government would try to negotiate voluntary agreements or covenants with the organisations. In other cases, it preferred regulation, financial incentives and information. One important financial incentive would be taxation of polluting products such as petrol. Another way to discourage the use of cars would be to abolish the tax deduction of commuting expenses for private cars (*reiskostenforfait*). Both measures would work like double-edged swords: as well as discouraging polluting behaviour they would raise money for other environmental policies.

More than any other aspects of the Plan, the proposal to abolish tax deductions for travel expenses met with fierce resistance from pressure groups and public opinion. It even caused the Liberal Party to end its coalition with the Christian Democrats and to topple the government. Unsurprisingly the environment figured prominently in the ensuing election campaign. Forty-five per cent of Dutch voters regarded it as the most important issue—more important than unemployment, health or any other problem. Yet only 34 per cent of the electorate was willing to forego tax deduction for travel expenses.

The new government—a coalition of Christian Democrats and Labour Party—tried to implement the National Environmental Policy Plan when it was approved by parliament in 1990. However, travel expenses remained tax deductible. Other measures to discourage the use of cars proved equally controversial—and if implemented at all, not very effective. More successful were voluntary agreements with industries, energy producers and construction companies to reduce or recycle waste and to combat the pollution of air and water. Fairly successful as well were policies concerning water management, waste separation and the conversion of cultivated land into 'natural' areas like forests or marshes. Some critics feel, however, that the Dutch have applied mainly 'end-of-pipe' solutions and avoided structural answers to the most difficult questions, such as water depletion (desiccation) and acidification, caused mainly by agriculture and traffic, the control of waste and the extinction of several species of plants and animals in the 'natural areas'.[3]

The National Environmental Policy Plan was evaluated and renewed by the government in 1993 and by parliament in 1994. By then, however, other political issues again began to 'crowd out' the environment: employment, social security and immigration were considered more important by voters in 1994 than pollution. A mere 5 per cent still gave priority to environmental problems. In the general elections of 1994, these problems did not play a significant part. After the elections, a coalition was formed between Liberals, Democrats and Labour, which intended to improve the infrastructure of the country by constructing tunnels, railway lines and expanding Amsterdam Airport (Schiphol): measures which would probably do more harm than good to the environment. Opposition from the environmental movement—as well as from many Not-In-My-Back-Yard groups—managed to slow down but rarely to prevent the realisation of these infrastructural projects. Even so, a few positive measures were also taken, for example to promote organic farming.

The Environmental Consciousness of the Dutch

This brief historical survey shows that the environmental consciousness of the government as well as the people of The Netherlands has been subject to considerable fluctuations. For most Dutch people environmental issues remain quite abstract; only a minority claims to have suffered from pollution, noise or other environmental problems.[4] Yet most people who do not suffer personally tend to express serious concern about the state of the environment, even when they give priority to other issues at a particular moment.

Even if the Dutch do not show much more concern about the environment than other nations in Western Europe, they seem more willing to pay for it.[5] Though this willingness may not always be evident in specific cases, such as when prices for petrol are raised, it is true that Dutch people spend a lot of money to protect the environment. Thus they donate more money to environmental organisations like Greenpeace or the World Wildlife Fund than most other Europeans. They are also more inclined to join them. The three main environmental organisations—the two just mentioned and the Society for the Preservation of Natural Monuments—claimed two million members in 1995, which is more than their counterparts in more populous countries like Germany or France.[6]

Membership need not imply activity, of course. At present, the Dutch are less inclined to engage in environmental action than the British or the Germans. In the past, however, the Dutch did take part in actions, too. Demonstrations against nuclear power attracted up to 40,000 people in the late seventies. These demonstrations have become superfluous, for the time being: as already noted plans for new power stations have been shelved and closure of the existing ones was approved by parliament in 1994.

From these data one may safely conclude that environmental consciousness is fairly widespread in The Netherlands, both among politicians and among average citizens. Several factors may explain this. Three have been mentioned already in passing:

— the density of the population, which adds to the intensity of most environmental problems;
— the fragility of the natural environment in The Netherlands, which dates from a distant past;
— the scarcity of many resources at the national level—especially oil—which may make people more aware of the limited quantity of these resources at a global level as well.

Three other factors may have contributed:

— the Calvinist tradition, which fosters thriftiness and frowns on waste of any kind; moreover, a Calvinist culture may be more sensitive to prophecies of doom as spread by the Club of Rome than more optimistic Catholic cultures;
— concern for the quality of life, which depends also on the environment, is characteristic of all affluent countries with advanced systems of social security like The Netherlands. This is true of the Scandinavian countries and Austria, as the American political scientist Ronald Inglehart has demonstrated;[7]
— the Dutch party system is quite open and accessible to new parties, thus allowing green parties to enter and exercise pressure on both public opinion and government.

The Dutch Party System: Shades of Green and Grey

The Netherlands hashad a multi-party system for more than a hundred years. Even before the introduction of an electoral system of pure proportional representation, four or five major parties and several minor ones would hold seats in parliament. A party needs to attract only 0.67 per cent of the popular vote in order to win a seat in the Lower Chamber of Parliament, the Dutch House of Commons.

In a fragmented and open system like this, one would expect Greens to enter parliament without any trouble. Yet this was not the case. When a green party was set up in 1983, it suffered failure after failure. In 1984 it failed to win a seat in the European Parliament, which requires 4 per cent of the popular vote. The Greens received 1.3 per cent. In 1986, 1989 and 1994 the party failed to win any seat in the Lower Chamber of Parliament. At the general elections of 1994, the Greens obtained hardly more votes than the Party for Environment and Law (*Partij voor Milieu en Recht*), a conservative green group of only 35 members founded in 1993. In 1995 the Greens did succeed in getting their candidate elected to the Upper Chamber (the Dutch Senate), but only through a clever alliance with several regionalist parties. Even at the local level, its gains have been rather modest—four seats in municipal councils, out of a total of more than 11,000 seats. With only 600 members, the Greens do not carry much weight in the environmental movement either. The latter tends to ignore them. When the Dutch Friends of the Earth (called *Milieudefensie*, Defence of the Environment, in The Netherlands) published a comparative analysis of election platforms in 1989 and again in 1994, they omitted the Greens.

The dire state of the Greens can be attributed to some extent to tactical errors and organisational weakness. However, the main factor is probably their late birth: other parties had already claimed the colour green, literally in their election posters, or ideologically in their election platforms. In the sixties, the Provo movement—mentioned above—called attention to air pollution by cars and industries. In the seventies, the Pacifist Socialist Party and the

Radical Party (founded by leftist Catholics in 1968) advocated windmills and solar energy instead of nuclear power, cheap public transport instead of more motorways, and selective rather than unlimited economic growth. In the eighties, these two parties co-operated and finally merged with the reformed Communist Party and a small Evangelical (Protestant) Party into Green Left (*GroenLinks*).

The Greens still resent the name of the Green Left. They resent even more the good relations between the Green Left and the environmental movement, especially *Milieudefensie*. For example Bram van Ojik, the present chairman of *Milieudefensie*, was Member of Parliament for the Green Left in 1994; the first president of the party, Marijke Vos, worked on the staff of the Dutch Friends of the Earth. Yet even the rather hostile Greens cannot deny that the Green Left devotes a lot of attention to environmental issues. Whereas it may not be a 'deep green' or ecocentrist party in ideological terms, its leaders, activists and rank-and-file members are without doubt concerned about the environment. In a membership survey held in 1992, 37 per cent of the Green Leftists agreed with the ecocentric statement 'If I have to choose between the survival of Nature or the survival of mankind, I opt for Nature': 56 per cent of surveyed members of The Greens agreed with this.[8] The Green Left's views about the economy reflect still the socialist legacy of the Pacifist Socialist Party and the Communist Party, but even here one can detect a growing tendency towards a 'third option', neither socialist nor capitalist. Thus a majority of delegates (56 per cent) surveyed at the 1995 party congress agreed with the statement 'Both the market and the planned economy have failed, we should look for another economic order, based on small-scale communities'.

At any rate, Green Left seems the greenest party in the Lower Chamber of Parliament at present. Quantitative as well as qualitative analyses of election platforms show this quite clearly. A quantitative comparison of the percentage of words devoted to environmental issues (in the widest sense) in party platforms can be interpreted as a very rough indication of environmental concern. Of course, some parties talk a lot about the environment without promising any specific policies. Hence a qualitative analysis is needed, too. In 1994, we selected five issues that seemed particularly controversial in the election campaign, such as: the expansion of Schiphol (Amsterdam Airport), closure of the remaining two nuclear power stations, transition from 'factory farming' to organic farming, and green taxes, especially on energy. In all issues, the Green Left took the greenest position. Moreover, when people in the national election study of 1994 were asked: 'which party has the best ideas about how to solve the pollution problem?', almost half (48 per cent) replied: the Green Left.

With 12,000 members, about 400 municipal councillors, 5 Members of Parliament in the Lower Chamber and 4 in the Senate, the Green Left carries more political weight than the Greens. In some cities like Amsterdam it took part in local government. Yet at the national level it has been and probably will be a permanent opposition party. Though its influence is limited, it has occasionally added items to the political agenda, for example the 'green tax' on fuel.

Much more influential, of course, are potential government parties such as the Christian Democratic Appeal (CDA) and Democrats 66 (D66). Both parties carry the colour green in election posters and other material. The Christian Democrats devoted even more words to environmental issues in their 1994 election platform than the Green Left. Voters were a little sceptical, however, as to whether they had better ideas to solve the pollution problem: only 9 per cent credited the CDA with the best ideas. The experts of Friends of the Earth (*Milieudefensie*) were even more sceptical about the CDA than the voters.[9] While the Christian Democrats talk a great deal about man's duties as steward of God's creation, they attach more importance to economic growth than to protection of the environment. With respect to the five most relevant environmental issues in 1994, they adopted a moderate green position in three cases and a grey one in two. They approved of the expansion of Schiphol and opposed the closure of nuclear power stations.

Democrats 66 seem slightly more justified in claiming the colour green than the CDA, both in the eyes of the voters and in those of Friends of the Earth. Since its foundation in 1966, D66 has emphasised typical post-materialist goals and values, such as the quality of life and protection of the environment, but also participatory democracy and constitutional reforms. Yet its pragmatist rejection of any ideology prevents a profound commitment to green policies. Thus in government the Democrats have often compromised their ecological ideals under pressure from economic interests. They have been consistent in their opposition to nuclear energy, but accepted the expansion of Schiphol.

In practice, the Dutch Labour Party (*Partij van de Arbeid*) looked almost as green or grey as D66. Like other social democrats, Labour supports sustainable development, out of solidarity with future generations. Earlier than sister parties elsewhere, the Dutch social democrats renounced nuclear power. Yet they also favour large infrastructural projects, whether railway lines or airports, in order to create more jobs and economic growth.

The Labour Party are certainly greener than the Liberals (*Volkspartij voor Vrijheid en Democratie* (VVD)) in The Netherlands. In 1994 only 6 per cent of Dutch voters believed the Liberals had the best ideas about pollution control, while Friends of the Earth awarded them the lowest score in their assessment. With respect to all five salient environmental questions in 1994, the Liberals consistently rejected the green alternative and opted for more highways, a larger Schiphol and more nuclear power, while opposing green taxes and organic agriculture. Thus the Liberal party met with criticism even from its own environmentalist wing. This still exists: some Liberals have been good ministers of environmental affairs and play leading roles in traditional environmental organisations like the World Wildlife Fund or the Society for the Preservation of Natural Monuments.

The three Protestant parties represented in the Dutch Parliament differ among themselves in denominational background, but also in 'greenness'. Most Dutch voters ignore their environmental programmes and associate them only with religious and moral issues. Yet the Reformed Political Federation (RPF) can certainly be considered an environmentalist party, if not 'green' in the ideological sense. It adheres to a theocentric world view, which is neither ecocentric nor anthropocentric, but which emphasises man's duty to respect God's creation. In many concrete questions the RPF adopted positions close to the Green Left, surprisingly: in favour of organic agriculture, green taxes, closure of nuclear power stations and support for public transport. Much more 'grey' is the pietistic Reformed Political Party, which caters more to farmers and fishermen than to environmental interests. The third Protestant party occupies a position between the other two.

None of the parties represented in the Dutch parliament can be considered completely 'grey': almost all insist on 'sustainable growth', reducing pollution, expanding natural areas and saving energy. An anti-environmentalist 'Car Party' was founded before the 1994 elections but failed even to muster the resources to participate in those elections—unlike its Swiss counterpart. And even the xenophobic Centre Democrats, who advocated lower rather than higher fuel prices, claim to be an environmentalist party and show great concern about animals. At the other end of the spectrum, Green Left is not 'deep green', in the sense of ecocentric ecologism, but rather 'shallow green'. Thus all Dutch parties represent shades of green and grey. This allows them to compete for green voters whenever environmental questions seem salient enough. Therefore, the Dutch party system may be more open to green ideas than many other systems, in spite of the relative weakness or even absence of a 'pure' green party.

After Autumn: Speculations on the Future

In the present season of Dutch politics, grey rather than green prevails. After the green spring of the early seventies, when every Dutch politician read *Limits to Growth*, and a green summer when the National Environmental Policy Plan reached maturity, autumn has come. Some environmental policies have borne fruit: the quality of rivers has been improved, SO_2 emissions have been reduced, nuclear power is almost abandoned, waste is separated and often recycled, and natural parks are 'developed'. Yet CO_2 emissions have increased, energy consumption remains high, eutrophication and desiccation of waters continue, exhaust fumes and manure pollute the air and acidify natural areas, animal and plant species disappear. At the same time, voters as well as politicians seem to have lost interest in environmental issues.

We may be heading for a cold and grey winter in The Netherlands. Yet as we all know, autumn leaves feed the earth, which will turn green again in spring. Environmental questions are bound to return to the political agenda, and both the Green Left and the Greens may gain more influence than they enjoy now. Yet for the time being, they have to content themselves with collecting acidifying autumn leaves.

Biographical Note

Paul Lucardie is Research Associate at the Documentation Centre of Dutch Political Parties at the University of Groningen in The Netherlands. His publications include *The Politics of Nature* (edited with Andrew Dobson), 1995.

Notes

1 Kees Aarts, 'National politieke problemen, partijcompetentie en stemgedrag', in: J. J. M. van Holsteyn and B. Niemöller, eds., *De Nederlandse kiezer 1994*, Leiden, DSWO Press, 1995, pp. 173–90.
2 Nico Nelissen, Rom Perenboom, Paul Peters and Vincent Peters, *De Nederlanders en hun milieu. Een onderzoek naar het milieubesef en het milieugedrag van vroeger en nu*, Zeist, Kerchebosch, 1987, p. 58.
3 For example Lucas Reijnders, 'Hebben we de milieuproblematiek enigszins onder controle?', in: Pieter van Driel, ed., *Terugtocht en vooruitgang*, Amsterdam, WBS, 1994, pp. 21–6; only slightly less critical are the authors of the *Environmental Performance Reviews: The Netherlands*, Paris, Organisation for Economic Co-operation and Development, 1995.
4 In 1985 24 per cent of the respondents in a national survey; see: Nico Nelissen, Rom Perenboom, Paul Peters and Vincent Peters, *op. cit.*, p. 100.
5 J. W. Becker, A. van der Broek, P. Dekker and M. Nas, *Publieke opinie en milieu*, Rijswijk, Sociaal Cultureel Planbureau, 1996, pp. 127–56.
6 Hein Anton van der Heijden, 'Politieke mogelijkhedenstructuur en de institutionalisering van de milieubeweging', *Acta Politica*, 1996, pp. 138–63.
7 See for example: Ronald Inglehart, 'Public Support for Environmental Protection: Objective Problems and Subjective Values in 43 Societies', *PS Political Science and Politics*, 1995, pp. 57–72.
8 For a more detailed comparison of the Green Left and the Greens, based on membership surveys, see: Paul Lucardie, Gerrit Voerman and Wijbrandt van Schuur, 'Different Shades of Green', *Environmental Politics*, 1993, pp. 40–62.
9 Jelle van der Meer et al., 'Op zoek naar idealen', *Milieudefensie*, 1994, pp. 13–23.

Prospects: The Parties and The Environment in the UK

NEIL CARTER

DOES the environment suffer because green issues have failed to ignite party political competition? Despite the rapid growth in media interest and public concern about the environment in the late 1980s, green issues have played only a peripheral role in British party politics during the 1990s. Since its impressive performance in the 1989 European election, the Green Party has again subsided into political obscurity. The three major political parties, after taking a crash course in environmentally friendly rhetoric, have since returned to 'politics as usual'—the economy, taxation, health, education, law and order, and Europe. Despite the Conservative Government's mediocre record on the environment, its opponents have been neither willing (Labour), nor able (Liberal Democrats), to turn it into an issue of party competition. The aims of this chapter are to explain the limited party politicisation of the environment and to speculate whether greater party competition over the environment would produce more progressive policy outcomes.

The Greening of the Major Parties?

Over the last decade the environment has moved higher up the policy agenda to a position where any government, irrespective of political hue, will be obliged to respond to a growing range of environmentally-related issues and problems. However, the party politicisation of the environment has been slow, uneven and incomplete. Parties seem to be most receptive to environmental pressure at the mid-term stage of the electoral cycle when public concern tends to be highest and leaders are more receptive to environmentalists within their parties.[1] The party politicisation process began with the establishment of fringe ecology groups, such as the Socialist Environmental Resource Association, within each party during the 1970s. It gained pace in the mid-1980s with the publication of policy documents by the SDP and Labour, but this was insufficient to turn the environment into a campaigning issue during the 1987 general election. Political interest resurfaced after a speech by Mrs Thatcher to the Royal Society in 1988 in which she seemed to accept many of the scientific arguments about the dangers posed by global environmental problems. The Prime Minister's apparent 'conversion' contributed to the spiralling media and public interest that prompted the unprecedented support for the Green Party in the 1989 European election. After that election each party sought to improve its own rather thin green credentials with a

Published by Blackwell Publishers, 108 Cowley Road, Oxford OX4 1JF, UK and 350 Main Street, Malden, MA 02148, USA

further flurry of documents: the publication of the government white paper on the environment, *This Common Inheritance*, was followed in quick succession by Labour and Liberal Democrat documents. But, again, the environment was conspicuous only by its absence during the 1992 general election campaign.[2] The mid-term pattern persisted with the publication during 1994 of the Government's *Sustainable Development: The UK Strategy*, and with policy documents from Labour, *In Trust For Tomorrow*, and the Liberal Democrats, *Agenda For Sustainability*, each representing a significant advance on previous efforts; although the environment was a non-issue in the 1994 European election.

The 1997 general election saw the three major parties paying greater lip-service to the environment. The Conservatives published a separate Green Manifesto, the Labour manifesto included numerous environmental initiatives and the Liberal Democrat manifesto devoted a three page section to its environmental proposals which Charles Secrett, Director of Friends of the Earth, described as 'the greenest manifesto ever from a mainstream party'. Yet after a brief flurry of interest early in the campaign, including a special press conference by John Gummer and Kenneth Clarke to launch the Conservative manifesto, the environment effectively vanished from the campaigns of all three parties and was ignored by the media.

The consistent disappearance of the environment at general elections suggests that, despite the plethora of policy documents, no party has fully embraced the green message. One obvious explanation for this reluctance, as the cyclical record suggests, is that electoral considerations have played an important part in shaping party responses to the environment. Opinion polls show that environmental issues have yet to play a significant role in determining voting preferences in any general election. Yet even if the major parties had wanted to turn the environment into a campaigning issue, it might have done them little good. Issue voting assumes that voters can distinguish between different party positions on a key issue. But opinion polls have consistently shown that when asked which political party's views on the environment came closest to their own, most respondents either do not know, or else they pick the Green Party. This has reduced the incentive to raise the profile of the environment, because that would simply be encouraging environmentally concerned voters to vote Green (although this argument ignores the ability of parties to shape the political agenda). Given the political sensitivity of some environmental issues—carbon taxes, road pricing, VAT on fuel—the two major parties seem to have concluded that they have more to lose by promoting a greener platform: better to play safe by learning a new greener rhetoric, appropriating some 'safe' green policies, but resisting the full-scale politicisation of the issue.

There are also broader political explanations for the reluctance to grasp the environmental nettle. These 'producerist' parties remain committed to economic policies and spending commitments that are dependent on continued economic growth, so they are understandably wary of green arguments about the dangers of unmediated growth. Parties may pay lip-service to sustainable development, but its central tenet—that environmental protection is a pre-condition of future (sustainable) growth rather than its antithesis—has failed to dissuade most politicians from the simple dichotomous belief in the economy *versus* environmental equation. Despite the new green rhetoric, the rationale for government intervention has not advanced much beyond Harold Wilson's classic formulation to the effect that 'the Polluters are powerful and organised . . . the Protectors, the anti-pollution lobbies, are less organised, less powerful. Therefore the community must step in to redress the balance'.[3] And, in practice, fearing the anti-competitive effects of tough environmental regulations and ecotaxes, successive governments have been reluctant to step in too often or too firmly. This wariness is encouraged by the representatives of producer interests—farmers, industrialists, trade unions—who exert powerful external and internal pressure on both parties not to accede to the demands of the environmental lobby. The effectiveness of these interest groups in penetrating to the heart of government was illustrated when Chris Patten's ambitious and radical plans, for *This Common Inheritance*, ran aground on the sands of Whitehall departmentalism. For the producer interest groups—farmers, industrialists, road-builders, car manufacturers, energy utilities—were ably represented in Cabinet by their 'sponsoring' departmental ministers.

The record of the Conservative Government over the last decade illustrates the limited commitment to environmental protection. There were some important developments under the Conservatives. For example, the *Environmental Protection Act 1990* established the Integrated Pollution Control framework, and the *Environment Act 1995* set up a new, cross-sectoral Environment Agency. For the first time, a comprehensive environmental policy based on the principles of sustainable development was laid out in *This Common Inheritance* and *Sustainable Development: the UK Strategy*. While progress in integrating environmental policy-making has been slow, it seems that the recent decision to scale-back the massive road-building programme arose partly because the Department of Environment was able to highlight the contradictions between sustainable development and current transport policy. Various small but significant decisions, such as altering planning policy advice to block the development of new out-of-town shopping centres, led many environmentalists to regard John Gummer as the 'greenest' Secretary of State for the Environment.

Yet, Gummer seemed to fight a lone battle within Cabinet. Despite the apparent commitment to integration, government policy remained largely reactive, piecemeal and tardy. In particular, European legislation and directives frequently forced the Government to act, notably to improve drinking and bathing water quality. It was international pressure arising from UN Agenda 21 strategy that pushed the Government into producing the Sustainable Development documents. The solitary domestic environmental commitment in the 1992 Conservative manifesto, to create an Environment Agency,

194

was twice postponed before it was eventually set up in April 1996. A litany of actions, including the commissioning of the THORP nuclear waste re-processing plant and reneging on the retrofitting of the flue gas desulphur-isation equipment to power stations, revealed the absence of strategic, integrated environmental planning.

Conservative Party enthusiasm for environmental protection is also tem-pered by its ideological distaste for excessive planning, intervention and regulation—all of which are central to sustainable development. The integra-tion of environmental considerations into policy making across different sectors requires planning. The development of recycling industries and new greener technologies will benefit from government intervention. Conservative deregulatory convictions created a climate in which rather than extend protective regulations, existing ones had to be defended. Despite expressly supporting the use of market mechanisms to achieve environmental aims, the only significant examples have been discriminatory taxation in favour of lead-free petrol and a landfill tax. For many eco-taxes, such as a carbon tax and road-pricing, represent a potentially unpopular market intervention.

Overall, the Conservatives have been reluctant environmentalists, willing to act when necessary, but prepared to ignore, delay and dilute their responses whenever possible. But will Labour be any different, given the low profile of environmental politics?

Labour has an ambivalent attitude towards the environment. Labour governments have a respectable record promoting public transport, improv-ing access to the countryside and regulating against pollution, and many party members are active environmentalists. But there is a long-standing suspicion that environmentalism is the preserve of the middle classes who, in Crosland's words, want to 'kick the ladder down behind them' by focusing on threats to the countryside while ignoring urban decay and the material needs of the working classes.

Labour does have a far-ranging and comprehensive policy document, *In Trust For Tomorrow*, which, if implemented, would represent significant progress towards sustainable development. The cornerstone of the strategy is an annual Sustainable Development Plan containing a range of targets and timetables for action. Various institutional reforms, including a new Environ-ment Audit Committee to monitor the implementation of the plan, would enhance co-ordination of environmental policy across sectors. In some policy areas Labour proposals are quite radical, including a major shift in transport investment from road-building to public transport, a massive home energy conservation programme, and the establishment of legally-enforceable envir-onmental rights to clean air and water, information and compensation for damage.

However, Labour's thinking has been least innovative in the key area of economic policy. In practice, Labour still tends to 'talk about the environment only when it is talking about the environment'. A proper concern for the environment requires an integrated approach that incorporates all policy areas. But when Gordon Brown talks about creating 'sustained economic growth', one suspects he is not too concerned whether this is 'ecologically' sustainable. Trade union fears about the employment implications of green policies are addressed by means of an environmental 'New Deal' that would create thousands of new jobs in industries such as recycling, energy efficiency and environmental protection. But the substantial economic restructuring required for sustainable development would still make frictional unemployment unavoidable, as jobs shift from polluting industries into labour-intensive sectors. Significantly, at the 1994 annual party conference, the powerful TGWU and AEEW voted against accepting *In Trust For Tomorrow*, and the GMB abstained, as union leaders expressed concern about jobs in the opencast mining, nuclear energy and road-building industries. Although these union suspicions have since been allayed, it is not hard to imagine unions joining with employers against a Labour Government in opposing tougher environmental regulations that impose costs on industry and potentially reduce competitiveness. The Treasury may also baulk at the sheer cost of some proposals, such as fitting FGD technology to power stations and modernising the railway network. Labour may not share the Conservative's ideological antipathy to planning and intervention, but it still faces similar electoral and political pressures to those constraining Conservative policymakers.

Suspicions about Labour's commitment have been fuelled by the lack of enthusiasm for green issues from the party leaders. Neil Kinnock was uninterested in the environment. John Smith was only a little more sympathetic. The jury is still out on Tony Blair. He got off to a poor start by replacing Chris Smith, a genuine environmentalist, with Frank Dobson, most definitely not an environmentalist. Joan Ruddock was given the 'green spokesperson' portfolio, but outside the shadow cabinet. This status was restored only because the 1996 shadow cabinet elections required Blair to give Michael Meacher a post, although he carried little influence with the leadership. Blair himself has shown little interest in the subject. He delayed making a keynote address on the environment until February 1996 when, in a speech that 'might have been given by Mr Gummer',[4] he displayed a marked reluctance to make any specific commitments. Although the new 'Clause 4' includes a specific reference to working for 'A healthy environment, which we protect, enhance and hold in trust for future generations', the *New Labour, New Life for Britain* policy statement included only around 200 words on the environment in a 10,000 document, all located in a couple of paragraphs dedicated to the issue, rather than integrated into all policy areas, such as the economy, taxation and employment. The election manifesto did to some extent rectify this weakness by emphasising the need for integration and scattering green initiatives throughout the document. However, the environment portfolio, still held by the out-of-favour Meacher, did not merit a seat in Blair's first Cabinet.

The Liberal Democrats are the 'greenest' of the three major parties. While many of their policies are similar to those of the Labour Party, where the two parties differ sharply is in the effort made by the Liberal Democrats to integrate environmental considerations into economic policy. In particular, the Liberal Democrats have adopted a more radical approach to eco-taxes, including an endorsement of a carbon tax—a proposal that frightens the two main parties. An Environmental Audit Group vets all party policy statements. The Liberal Democrats have also been more willing to make the environment a central plank of their electoral strategy, notably in 1992 when it was promoted as one of the three key campaigning issues. Whilst the success of the Greens in stealing centre voters in 1989 obviously spurred the Liberal Democrats down this path, the party is also free of some of the pressures constraining the Conservatives and Labour. As an outsider in the electoral contest, the Liberal Democrats have least to lose; indeed, although a carbon tax could be very unpopular in their rural and suburban strongholds. The Liberal Democrats are less tied into producer interests than the other parties. But whilst the party has willingly seized the initiative, it has failed to alter the mainstream agenda: its environmental message is ignored by the media, and opinion polls show that the public does not regard the Liberal Democrats as any greener than the Labour Party.

The Green Party

In the absence of a genuine party political cleavage on environmental issues, does the Green Party offer a possible alternative? The Greens have had little tangible impact on British politics, having failed to gain representation in either national or European elections[5] with a miserable share of the vote (see Table 1). The one exception was the 1989 European Parliamentary election, when Green candidates polled an unprecedented 14.9 per cent of the vote, saved every deposit, came second ahead of Labour in six constituencies, and

Table 1 Green Party Election Results 1974–97

Election		Number of candidates	% of vote in seats contested
General	1974	5	1.8
General	1974	4	0.7
General	1979	53	1.5
European	1979	3	3.7
General	1983	106	1.0
European	1984	16	2.6
General	1987	133	1.3
European	1989	79	14.9
General	1992	253	1.3
European	1994	84	3.2
General	1997	95	1.4

beat the Social and Liberal Democrats in all but one constituency. Although the electoral system prevented the Greens from securing any MEPs, it seemed that the long awaited breakthrough was just around the corner. Membership doubled to 20,000 within a year. For the first time, mainstream parties acknowledged the Greens as a serious political force. Leading Greens, notably Jonathan Porritt and Sara Parkin, were given unprecedented media access. Subsequently, however, the Greens have returned to their former obscurity: their vote has collapsed, membership has fallen to below 4,000 and many leading figures have withdrawn from active politics. What went wrong? What are the future prospects for green politics in Britain?

It is necessary to set the sudden success of the Green Party against its previous failure. Small parties find it difficult to break into the first-past-the-post British electoral system in which individual constituency contests are usually dominated by the major parties. Electors are unwilling to 'waste' their votes on a party with little chance of winning a seat. Only where a party can concentrate its vote geographically, as with the Welsh and Scottish nationalists, is there a chance of gaining representation; but the Greens have been unable to establish any regional base. Green parties elsewhere have benefited from electoral systems based on proportional representation, usually with a threshold whereby once a party receives, for example, 5 per cent of the vote in Germany, it automatically secures parliamentary representation. In Britain, there is no state funding for political parties, so small parties are further penalised by the need to pay a £500 deposit for each candidate in a general election. Lost deposits in 1992 cost the party £126,500, a large bill for a small party.

The Green Party has been weakened by the refusal of the large environmental pressure groups to build close links with it. These groups make a virtue of their non-partisan status, believing they will exercise most influence by direct lobbying of civil servants and politicians. The environmental lobby sees little to gain from working with a Green Party with no MPs; indeed, such partisanship might close the doors to government and risk alienating its membership.

Unlike its sister parties on the continent, the Green Party has been unable to broaden its political base by building links with new social movements, such as the anti-nuclear and peace movements. Instead, the closure of the electoral system combined with the historical openness of the Labour Party to dissident social movements such as CND, feminism or anti-poll tax, has encouraged new social movements to work with, or within, the Labour Party. The absence of such links has encouraged the perception of the Green Party as a narrow, single issue party.

Thus the closure of the 'political opportunity structure' explains the failure of the Green Party before 1989, just as the transitory opening up of this opportunity structure accounts for its freak performance in the European election. That success had very little to do with the efforts of the party itself: rather, the Greens piggy-backed on the growing public interest in the environment and benefited from the specific political context in which the election was fought.

Public knowledge about environmental issues grew rapidly during the late 1980s, spurred on by a series of 'eco-disasters', such as Chernobyl, and the emergence of new global issues, such as the greenhouse effect, to more parochial concerns such as the privatisation of the water industry, salmonella in eggs and dying seals in the North Sea. The widespread public concern was manifested in several ways during 1988–9: a rapid growth in the membership of environmental pressure groups; the emergence of green consumerism; extensive coverage of environmental issues by the media; and the gradual politicisation of the major political parties. In the run-up to the European election, opinion polls reported that around a quarter of respondents regarded the environment as one of the two most important issues facing the country.

Political conditions were also ideal for the Greens. The public clearly treats European elections differently from domestic general elections. They provide an opportunity to express discontent with the government, either by not bothering to vote (turnout was only 36 per cent) or by casting a protest vote without running the risk of putting an alternative party into power. In 1989, the Conservatives were suffering from typical mid-term unpopularity, exacerbated by the recently introduced poll tax. The Labour Party, although regaining popularity under the modernisation programme of Neil Kinnock, was still widely distrusted. Crucially, the centre ground was also unattractive due to the bitter feuding between the former Alliance partners, the SDP and the SLD, following the 1987 general election. None of the major parties offered much to attract those voters concerned by environmental issues. In contrast, the Green Party had an unambiguously 'green' identity at a time when there was unprecedented interest in environmental issues.

This interpretation is supported by analysis of the results: one survey showed that just 7 per cent of Green voters had previously voted Green, whereas 25 per cent had voted Conservative, 27 per cent SDP/SLD, 19 per cent Labour and 15 per cent abstained. Moreover, only 42 per cent of those voting Green said they would vote Green again and just 28 per cent felt 'close' to the party.[6] Thus the green vote in 1989 was predominantly from the unpopular centre-right parties and very unstable, which lends support to the protest vote thesis. The centre-right character of this support also suggested the continuing capacity of Labour to gather in the left-wing environmental vote. Many voters probably saw the Green Party more as a 'non-partisan' single issue pressure group than a conventional political party. Thus the Green vote was both a positive expression of concern about the environment and a protest vote.

Success was short-lived. Opinion polls show Green support falling dramatically in late 1990 to level out at around 2 per cent in 1991, where it has hovered ever since. Subsequent elections have been a disaster for the party (see Table 1). In the 1992 general election, a record 253 candidates secured a total vote nationwide of just 173,008—an average of 1.3 per cent in the constituencies contested, no better than its 1987 performance. In the run-up to the 1997 election, some activists argued that the Green Party should conserve its limited finances by not contesting the election. In the event, a much-reduced list of 95 candidates attracted a feeble 64,021 votes.

Just as its strong showing in 1989 had little to do with its own efforts, so the Green Party was largely powerless to prevent its subsequent rapid decline. The external conditions that attracted support in 1989—the high level of public concern about the environment combined with the particular balance of political competition—had disappeared long before the 1992 general election. The decline in the political saliency of the environment was dramatic: from a peak in July 1989 when 35 per cent of the public regarded the environment as one of the two most important issues facing the country, it went into free fall in late 1990, levelling out at approximately 6 per cent in early 1991. The environment was crowded out by growing public concern about traditional material issues—the poll tax, the health service and the deepening recession—which contributed to the ousting of Mrs Thatcher as Prime Minster. The 1992 general election was a two-horse race in a campaign dominated by economic issues; the environment and the Green Party were simply ignored.

While events beyond its control conspired to remove the environment from the political agenda and changed the nature of political competition, the Green Party probably contributed to its own downfall by engaging in bitter internal factionalism over organisational and strategic issues. Of course, electoral catastrophe would not have been avoided if the Greens had presented a united front to the world, with a moderate programme and an articulate and persuasive leadership. But the rapid haemorrhaging of membership might have been prevented, for in a small, impoverished party, activists are the lifeblood that help counteract the lack of financial resources.

The disastrous performance in the 1992 election probably represented its last serious attempt to secure parliamentary representation for the foreseeable future, at least under the existing electoral system. Subsequently, the party has focused on grassroots politics. The Greens hope to build a strong local base through active community politics,which might become the launching pad for electoral success in national or European elections under a reformed electoral system. Currently, there are around 25 green councillors, with small concentrations in Oxford and Stroud. The party has also engaged more openly in extra-parliamentary activities, building links with groups involved in direct action, such as the anti-roads protests and housing squats. This alternative strategy may make the party even more marginal in electoral terms,so that, until electoral reform, parliamentary green politics will be left in the hands of the major parties.

Pressures from Civil Society

If the Green Party is unlikely to succeed in raising the profile of environmental politics, can other forces in civil society make a greater contribution? The dramatic growth of non-violent, green direct action during the 1990s, notably against road-building and veal exports, has had a broad impact on the nature of UK environmental politics. The flourishing ALARM UK network of some 200 local anti-roads groups represents the most extensive and successful form of protest. Typically, individual campaigns, as at Newbury or Twyford Down, have consisted of an alliance of two groupings: local residents and full-time activists from the green counter-culture movement, or 'ecowarriors'. The involvement of local residents, fuelled by NIMBY anger, reflects a frustration with the political system that offers few legitimate means of opposing a road scheme. However, the ecowarriors represent a deeper alienation, particularly among young people, from mainstream parties and political institutions. The ecowarriors are drawn from a number of groups—Earth First!, the Donga Tribe, travellers, ravers, some Green Party members—and have managed to co-ordinate protest activity around the country and across a number of other issues, such as veal crates, the Criminal Justice Bill, open cast mining and 'Reclaim the Streets' actions. With the slowing down of the road-building programme, the ability of this radical movement to survive depends on its ability to make the linkages to these related issues.

Whilst direct action has secured no formal political influence and the protests are unlikely to be harnessed into effective political activity beyond single issue protests, the broader impact on environmental politics is significant. On the specific issue of roads, whilst several factors contribute to the decision to scale back the road-building programme, it is hard not to conclude that the direct action protests played some part in this U-turn. The sight of protestors up trees, or lying down in front of bulldozers, raised the media profile of the issue. The sheer cost of security and delays in major road-building schemes has also escalated to the extent that some road construction companies have refrained from tendering for new contracts.

The emergence of this new radical ecological movement also suggests that the ability of the British political system to accommodate environmental protest is weakening. In the absence of a successful Green Party, the established pressure groups have carried the burden of the environmental movement. With some 4–5 million members of groups, there is clearly enormous potential for environmental activism in Britain. The established groups have secured limited access to official policy-making processes across a range of issues, especially in areas not dominated by producer groups. But the price of this limited access to decision-makers is that the groups are constrained not to participate in action that might compromise that access. Thus even once radical groups, such as Friends of the Earth (FoE) and Greenpeace have increasingly tried to be law-abiding. The heavy emphasis on lobbying and fund-raising has resulted in members being discouraged from engaging in protest activity.

The direct action protests have posed a serious challenge to this strategy. Environmental activists, frustrated with the hierarchical and reformist approach of the established groups, have increasingly turned to the new radical fringe groups to pursue their political ends. Consequently, Green peace and FoE have become more flexible about engaging in direct action. In many anti-road campaigns, FoE worked alongside the radical group: the former using the official system, such as the courts; the latter taking direct action.

Another forum in which a range of pressure groups work together in Real World. This non-partisan umbrella group, representing a coalition of 32 campaigning and voluntary sector organisations, was launched in April 1996 with the aim of pushing the issues of sustainable development, social justice, and democratic renewal up the political agenda in the run up to the general election. Real World is interesting, because it draws together a disparate array of simple issue groups—including Oxfam, Save the Children, FoE and ALARM UK—that normally work within narrow issue areas.They now recognise the interlocking nature of these problems, such that success in one area represents progress for all. Although it was unable to make the environment an election issue in 1997, its longer term effectiveness may lie in its ability to persuade the new Government to address these issues.

The Future of Environmental Politics?

Over the last decade, despite its low electoral saliency, various domestic and international pressures have pushed the environment onto the political agenda. These pressures remain and many environmental problems seem likely to increase in intensity. Two obviously complex and contentious issues that will need addressing are: the conflict between countryside protection and the need for new housing, and that between road congestion resulting from growing car ownership and the need to improve the urban quality of life by investing in a crumbling public transport system and by reducing worsening air pollution. Solutions will increasingly need to be addressed within the framework of the emerging sustainable development process. Much will depend on the genuine commitment of the Government to that process, but the low profile of environmental politics to date suggests that Labour in power will be an unenthusiastic environmentalist, and may not implement many of its pre-election environmental commitments.

However, there are several factors that might encourage New Labour to seize the environmental initiative. There may be political gains to be made from projecting its green credentials. Young voters in particular express concern about environmental issues. This cohort may feel alienated from the existing political system, but the emergence of radical green direct action shows that political consciousness is not necessarily absent. Labour strategists may perceive it strategically profitable to raise the profile of environmental politics by projecting the party as significantly greener than the Conservatives, to retain the youth vote which it attracted in the general election. More broadly, some of the Labour's environmental proposals may encourage positive perceptions of government. For example, the implementation of a substantial energy efficiency programme that created jobs, produced observable changes in homes and cut energy bills—as well as reducing environmental damage—might prove a populist success. A positive environmental agenda may also be attractive to a party membership that may become frustrated by the constraints on Labour in other policy areas.

The presence of several enthusiastic environmentalists within the leadership, notably Chris Smith, Ann Taylor and Robin Cook, may also make the Government generally more favourable to adopting a greener image and, in practice, more accessible to the green lobby. With the balance of power in the party shifting away from the unions in general, and from manufacturing unions in particular, traditional objections to environmental policies may carry less force than in the past. Several unions are aware of the job potential in environmental technologies and recycling industries and have lobbied for such policies. The pre-election donation to the party of £1 million from the Political Animal Lobby—because of Labour's commitment to a free vote in Parliament on a bill to ban hunting—symbolises the changing nature of the Labour Party in which one of the remaining sources of trade union influence, its control of the party purse-strings, is diminished by the growing importance of new sources of funding.

At a more abstract ideological level, the environment has been neglected in the current intellectual debate about the future direction of the Labour Party. Yet in a political climate where the market is in ascendancy and public ownership unpopular, the environment offers an opportunity for Labour to develop a positive role for the state. Environmental protection under Labour would involve active government planning and intervention in the economy through, for example, a national energy efficiency programme, the promotion of greener industries using clean technologies and recycling waste products, and tougher regulations on polluting industries. In short, the environment offers a way of re-legitimising the state.

In the immediate aftermath of Labour's remarkable general election victory, there were signs that the new Government was giving some attention to the environment. Admittedly, the exclusion of Michael Meacher from the Cabinet suggested that there will be no one arguing the environmental case at the heart of the Government, whereas Transport, which historically has played the role of advocate for the road lobby, will be represented at Cabinet by Gavin Strang. However, both departments have been incorporated within the new 'super-ministry' covering Transport, Environment and the Regions, headed by the Deputy Prime Minister, John Prescott, who stated that he wants to encourage integration between two departments 'parts of which have not been speaking to each other for years'. Transport issues are central to any sustainable environmental programme and with Prescott declaring that he wants 'to improve public transport and make it more attractive so that people will use their cars less', it seems likely that many of Labour's radical proposals in this area will be implemented. It also makes administrative sense to tie transport issues more closely to the local government arena where many innovative Agenda 21 strategies have been developed in recent years. There was sporadic evidence that environmental considerations may inform other policy areas. In his first interview as Foreign Secretary, Robin Cook declared that one of the new Government's four foreign policy priorities would be 'to place human rights and environment at the centre of European policy', which fits Labour's aim of reforming the Common Agricultural Programme. However, not all Cabinet ministers share Cook's long-standing concern for environmental protection. It will be some time before we can judge whether the worthy manifesto declaration that the environment is 'at the heart of policy-making' and 'not an add-on extra' has any substance.

It has been argued that if Labour were to turn the environment into an issue of genuine party competition, with a much higher profile, then this could only benefit the environment through a more radical policy programme, but it could also have clear political advantages for Labour. If Labour refuses the environmental challenge in the short-term, then, longer term, the main hope for environmentalists is electoral reform. Of course, the prospect of electoral reform may, in itself, encourage Labour to raise the profile of environmental politics. Proportional representation would provide a genuine opportunity for the environment to emerge as a key political cleavage in British politics, either via the Green Party or a Liberal Democrat Party with a fairer representation in Parliament. Labour may seek to pre-empt such a development by acquiring a genuine green tinge. Either way, many environmentalists will have one overriding objective for the Labour Government: to win the promised referendum on electoral reform.

Biographical Note

Neil Carter is a lecturer in politics at the University of York and author of several articles and chapters on British political parties and on environmental politics.

Notes

1 Andrew Flynn and Philip Lowe, 'The Conservative Party and the Environment' in W. Rüdig, *Green Politics Two*, Edinburgh University Press, 1992.
2 See Neil Carter, 'Whatever Happened to the Environment? The British General Election of 1992', *Environmental Politics*, Vol. 1, 1992, pp. 442–8.
3 Harold Wilson, *New Society*, 5 February 1970.
4 *ENDs Report* No. 252, February 1996, p. 3.
5 Cynog Dafis was elected MP on a joint Plaid Cymru/Green ticket in 1992, although he has since dropped the Green association.

6 Wolfgang Rüdig and Mark Franklin, 'Green Prospects: The Future of Green Parties in Britain, France and Germany' in *Green Politics Two, op. cit.*

Index

acid rain 132–3
Advisory Committee on Business and the Environment (ACBE) 102
agriculture 110, 125
air quality 134–6
ALARM UK 201
Arendt, H. 25
Aristotle 157

baby milk 125–6
Beck, U. 18–33
  critique of risk theory 34–44
Benn, G. 31
Berking, H. 153–4
Biotechnology and Biological Sciences Research Council (BBSRC) 124–8
Blair, T. 196
Brandt, W. 174
Brent Spar 114–15, 131
Brown, G. 196
Brundtland Report 3, 71, 87, 138, 148, 152
BSE 18, 22, 27, 39
  politics of 39–40, 115–16
business community 67–9, 80
  and tax reform 103–4, 108

Cambridge Econometrics 99
capitalism 38–42, 43
  and ecological modernisation 75–6
Carson, R. 68
Central Europe 76
Church Action on Poverty 72
civil service 112–13
civil society 119–21
Clarke, K. 99, 101–2, 193
class divisions 37, 42–4
  see also 'poverty lobby'; social justice
clean technologies 75, 78–9, 80–1
Climate Change Conference 104
Club of Rome 184
Commission on Social Justice 102
community-based partnerships 140–6
Conservative Party 193, 194–5

consultation 124–9, 131–3, 135–6
  in local government 140–6
consumer demands 10–11, 91–3, 106
Consumers in Europe 106
consumers' organisations 106, 107–8
consumption 47–60
  reduction 48
  as social force 51
Cook, R. 204
cost-benefit analysis 117, 133–4

Daily Telegraph 66
'deep ecology' 71
democratic deficit 55–7
Denmark 101, 128
Department of the Environment 111–12, 132, 134–6
Deutsches Institut für Wirtschaftung 100
direct action campaigns 65–7, 117, 201–2
DLO 127–8
domestic fuel tax 105, 107, 134
downshifting 50–1

Eastern Europe 76
ecological modernisation 8–10, 74–85
economic development 74–5
Edmonds, J. 107
efficiency measures 48
elected councillors 146
employment 100, 102, 106–7
energy consumption 47
Engels, F. 43
Engholm, B. 177
English Nature 88
environmental efficiency 74–5
Environmental Industries Commission 104
environmental movements 44–5
environmental tax reform 54–6, 98–108
European Environment Agency 101
European Union 78
  Fifth Action Programme 98–9
  taxation 98, 100, 103
Ewald, F. 23

© The Political Quarterly Publishing Co. Ltd. 1997
Published by Blackwell Publishers, 108 Cowley Road, Oxford OX4 1JF, UK and 350 Main Street, Malden, MA 02148, USA